The
Island *in*
Imagination
and # Experience

The Island in Imagination and Experience

Barry Smith

Published by Saraband,
Digital World Centre,
1 Lowry Plaza,
The Quays,
Salford M50 3UB,
United Kingdom

www.saraband.net

ISBN: 9781910192795
ebook: 9781910192801

Printed and bound in Great Britain by Clays Ltd, St Ives plc.

Editor: Craig Hillsley
Illustrator: Pete Smith, Picturemaps
Map designer: Ruaridh Cunliffe (map outlines © d-maps.com)

Contents

The Pacific Ocean

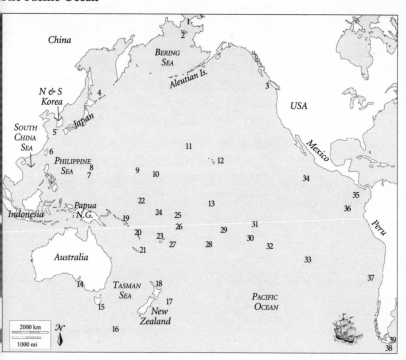

North Pacific
1 Big & Little Diomede
2 St Lawrence
3 Vancouver
4 Kurile
5 Jeju
6 Taiwan (& Orchid)

Mid Pacific & Micronesia
7 Guam
8 Mariana
9 Bikini
10 Marshall Islands
11 Midway
12 Hawaii
13 Kiribati

South Pacific
14 Kangaroo
15 Tasmania
16 Macquarie
17 Chatham
18 Great & Little Barrier
19 Solomon Islands
20 Vanuatu
21 New Caledonia
22 Nauru
23 Fiji, Taveuni
24 Tuvalu
25 Tokelau
26 Samoa
27 Tonga
28 Cook Islands

29 Leeward & Windward
   Islands
30 Tuamotu Islands
31 Marquesas Islands
32 Pitcairn
33 Easter Island

Central & South America
34 Clipperton
35 Cocos Island
36 Galapagos Islands
37 Robinson Crusoe (Más a
   Tierra)
38 Cape Horn, Nueva &
   Diego Ramirez
39 Tierra del Fuego

North Atlantic Ocean

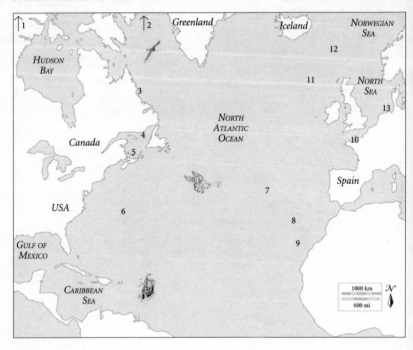

Caribbean Sea and Central America

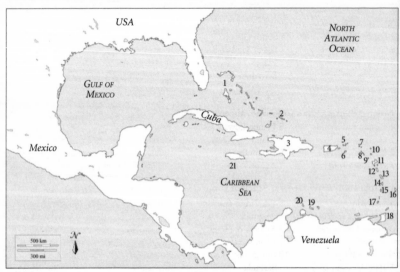

South Atlantic Ocean

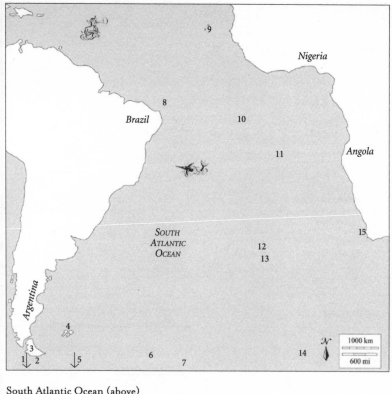

South Atlantic Ocean

Nigeria

Brazil

Angola

SOUTH
ATLANTIC
OCEAN

Argentina

1000 km
600 mi

Top: **Mediterranean Sea;** *Below:* **Indian Ocean;** *Opposite:* **British Isles**

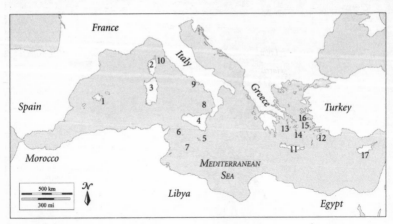

1 Mallorca	6 Pantelleria	10 Elba,	14 Santorini,
2 Corsica	7 Lampedusa	Pianosa	Amorgos
3 Sardinia	8 Stromboli,	11 Crete	15 Patmos, Leros
4 Sicily	Lipari	12 Rhodes	16 Agathonisi
5 Malta	9 Ponza, Procida	13 Seriphos	17 Cyprus

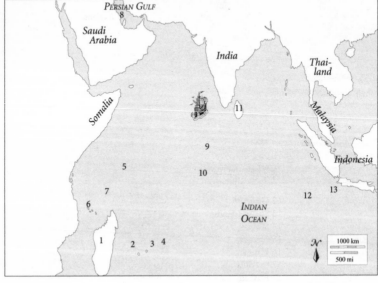

1 Madagascar	5 Seychelles	9 Maldives	12 Cocos
2 Réunion	6 Comoros	10 Chagos Islands,	(Keeling)
3 Mauritius	7 Aldabra	Diego Garcia	13 Christmas
4 Rodrigues	8 Bahrain	11 Sri Lanka	

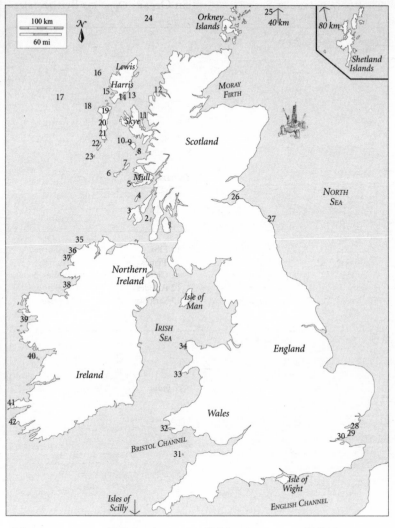

1 Arran	12 Gruinard	23 Mingulay	33 Bardsey
2 Gigha	13 Shiant Islands	24 North Rona	34 Anglesey
3 Islay	14 Scalpay	25 Fair Isle	35 Tory
4 Jura	15 Taransay	26 Inchcolm	36 Gola
5 Iona	16 Flannan	27 Farne Islands	37 Arranmore
6 Tiree	17 St Kilda	28 Mersea	38 Inishmurray
7 Coll	18 Haskeir	29 Foulness	39 Clare
8 Eigg	19 North Uist	30 Canvey	40 Aran Islands
9 Rum	20 Benbecula	31 Lundy	41 Blasket Islands
10 Canna	21 South Uist	32 Skomer,	42 Skellig Michael
11 Raasay, S. Rona	22 Barra	Skokholm	

Introduction

The Island

Look up at the Milky Way on a clear, dark night. Try to focus on the furthest of the furthest stars, the ones that seem to flicker into and out of existence. Consider what a light year is – a little less than 9.5 trillion kilometres – and how many light years away these stars are. Focus again on those furthest stars, and remind yourself that this is merely an infinitely minute bit of deep space. At this point, one's imagination pops like a failing light bulb.

Now look at a decent-sized globe of the world. Spin it around to the Pacific. This is where you will find a good proportion of the world's half-a-million islands. Turn the globe 180 degrees to remind yourself of scale, and then spin it back again. Trace the strings of pearls that make up the far-flung archipelagos – the space is vast, but it is just about imaginable. And so it is with small islands: not like stars that fall off the edge of one's imagination, but tiny pieces of land, the existence of which imagination can just about hang on to.

In the British Isles alone there are reckoned to be some 6,289 islands anchored near and far around its shores. The figure is approximate – I have relied on others more patient than I to count them – and even now ocean tides and currents, rivers and the wind are combining forces in making new islands, just as they are sweeping others away.

Born into this island nation, I was destined to live my early years in Essex, which, unbeknown to me then and a surprise to me even now, has more islands than any other county in England. The Isle of Dogs and Canvey Island were places of significance in my childhood geography, the former because it is close to where my father was born, and the latter because of catastrophic floods in 1953 when

I was five years old. Indeed, that North Sea flood, together with the Coronation and the first ascent of Mount Everest, constitute my earliest memories; but as real islands, Canvey, Foulness, Mersea and the like barely merited my attention, separated as they are from the Essex mainland by sandflats, narrow channels and barely discernible creeks. And neither could the Isle of Dogs be held accountable for firing the imagination of a young islomane, being a former island where the hand of mankind has reversed geography to make a peninsula bounded on three sides by a meander loop of the River Thames.

So my first real island experience was one that captured my imagination far more than any land surrounded by meander loops or marshy creeks could achieve; and, as with many young people, it was an experience infused with an adventure that even sixty years later can raise my pulse. When I was at primary school it was just possible during lunch break to sneak out-of-bounds, cross a field into a woodland, and look upon a lake with an island in the middle. It was a tangle of wild nature far more verdant than where we stood. And it was inaccessible and protective of its hidden treasures. Only my elder brother had the courage to lead me there, and I believed that nobody else had seen it. It was *our* island, but how could we claim it if we could not make a landing and hoist a flag?

Then one winter the annual disappointment of only a few days of snow was reversed, and the lake was frozen. We did not dash onto the ice, as we were surely tempted to, for we were of an age where even faint memories of parental admonishments conjured up fear. No, we were careful. We tiptoed from the shore a couple of steps, then tiptoed back, reversing the procedure and gaining courage all the time. Soon we were a good few yards out, testing the ice by crouching up and down, then by jumping up and down. And then we were walking matter-of-factly towards the island, our progress only impeded when faith exceeded friction and we fell unceremoniously on our backsides. We were nearly there; the key to the island's secrets was within our grasp. We were at the point of leaping ashore when the woodland echoed with the awful sound of cracking ice, and I felt my altitude drop an inch or two. The horror – transfixed by fear in the knowledge that any movement from either of us could be our last – could have continued for eternity had not the distant sound of the school bell presented us with the even greater fear of being late for afternoon classes. We slithered and crawled on all-fours to the bank, and scrambled blindly through the thickest part of the wood, until we were free to run ever so easily towards the far buildings.

I was frightened, dirty and scratched, but my tears were not because of this. Nor were they because I would soon be subjected to teacherly interrogation concerning my condition. No, I was crying because I knew that the secrets of the island, which had been so close, would never be revealed to me. And I was right. More than fifty years later, my Essex adventure still lingered and rankled faintly yet insistently in my memory. During those years I had made the final jump onto so many distant islands in far-flung reaches of the globe: summer by summer exploring the Scottish Hebrides, culminating in a circumnavigation of the Outer Hebrides by kayak; undergraduate days island-hopping in the Mediterranean and scaling its active volcanoes, and then undertaking an expedition to one of the smaller islands in the West Indian Grenadines; kayaking to the islands of Atlantic Wales, and along the Dalmatian coast of what was then Yugoslavia; expeditions to the archipelago south of Tierra del Fuego and in south-east Alaska; and sailing voyages throughout British Columbian coastal waters.

But I still needed to return to the first island that had beckoned me as a mere child; that had tempted me to cross its protective icy fringe. Perhaps once trodden securely underfoot, I could close a sometimes exhausting chapter in my life during which islomania had become as much a compulsion as a desire. But trying to retrace one's steps through life can be more perilous than the original journey and, as this book should make clear, islomania is one of the most difficult compulsions to be cured of. In the event, the mental map that I had carried for fifty years was no help in navigating the new reality of a wholly transformed landscape: there was a scrap of a field colonised by rank weeds, there was no brambled wood, no lake and so no verdant island. There was just urban development, just a large regional hospital and all the paraphernalia that goes with that.

This fragment of autobiography provides a foundation stone for an exploration of the tenacious grasp that small islands have on the imagination, and the kind of human experiences that islands inspire. My school island harboured hidden secrets, secrets that once ashore would surely be easily discovered. But even though it was only a stone's throw away, it felt inaccessible and quite possibly unattainable. It was *my* island, and like me it was probably vulnerable; and indeed its life was short-lived, ephemeral like so many of its fellows.

The Island

Enisling water pressing insistently upon one's conscious-ness creates the impression that an island may be viewed at a glance, its fine detail explored in a day and so made intimate. Here is somewhere one may soon become master, a domain where inaccessibility and remoteness render it free from mainland influence. It is perhaps not surprising then that this is a place where some penitents will volunteer to escape from the temptations of the wider world, just as it is a place where some impenitents will be deposited for the sin of failing to resist different temptations. Castaways and maroons condemned to, at best, the horror of isolation, are probably less numerous in maritime history than voluntary maroons and beachcombers strangely attracted to the life of Robinson Crusoe. Small islands harbour the belief that here one may find a paradise that has been lost on the mainland, a place where one's utopic intentions are more likely to come to fruition, and a natural laboratory for testing all manner of brave new world ideas and nefarious devices that would be unacceptable on continental shores.

It is within this farrago of ideas that we will journey, intent on disentangling the often paradoxical qualities of island imagination and experience, and along the way we will start making sense of the insistent grip the island exerts on the human psyche.

Chapter One

Lost in Space, Lost in Time
Islands and the Imagination

The Island

There are about half a million islands in the world, most of them are small, and if we discount continental Australia and Greenland, they constitute little more than five per cent of global landmass. The Pacific alone contains some 23,000 islands, and one archipelago – Tuamotu – spans an area about the size of Western Europe.

Whilst islands are often clustered in far-flung chains, there are also dizzying amounts of space between them, especially in the high latitudes of the South Atlantic and in the southern and south-eastern Pacific, where small island groups are often extremely isolated. Indeed, the first European to enter the Pacific, Ferdinand Magellan in the early sixteenth century, sailed about 13,000 miles from Cape Horn to the Philippines and encountered only two islands which, despite the considerable discomforts of his own enormous voyage, he named the Unfortunate Islands.

My interest is in small islands and I do not have strong opinions about how these are defined. For the purposes of analysing island development issues, economists exercise considerable latitude in defining anything between Cuba (110,000 square kilometres) and Nauru (twenty-one square kilometres) as "small". An islomane – somebody who claims to love islands – may assert that an island is only small if the enisling sea impinges upon one's senses at all points of land. My focus lies somewhere between the two, probably nearer to the islomane and Nauru than the economist and Cuba; but smallness, like remoteness and proximity, is relative and contextual, and so it is in my discussion of how islands are imagined and experienced. [1]

That islands tend to be small, that they are sometimes clustered in large groups covering vast areas, and that they are often very remote from mainlands – these are well understood objective physical qualities, and it is appropriate that scientific enquiry enables generalisations to be made concerning islands and their spatial relationships with each other and the mainland. But how we *interpret* these physical traits and give them meaning is significantly dependent on the cultural, social and economic lifeworld of the perceiver: consequently, we attribute qualities to an environment that it does not intrinsically possess. We bring to bear a shared cultural memory and individual experience and imagination when we interpret landscape, so that there are in fact many versions of the same scene. [2]

An island may be considered *remote* by virtue of its distance from the mainland or from larger islands, and this is something expressed objectively. But it is *isolated* because that is what it *feels* like subjectively. Even a thin channel of water is an effective isolator, and more than anything else it is the effect of surrounding water on the psyche that gives islands their uniqueness. If one doubts this, one only has to look at the New York City archipelago which, including Long, Manhattan and Staten, consists of about forty islands. Most of these are within close, clear sight of the mainland with which, in appearance and history, they share very little. Roosevelt Island on the East River blends in with Queens and Manhattan on either side, but it is only separated by about 200 metres, and even here its use for an asylum, a workhouse, a prison and a smallpox hospital suggests a sense of isolation.

Elsewhere in the archipelago, enislement has fostered a strong sense of isolation and enabled activities to be concentrated in a manner unacceptable on the mainland.

The list is long: Rikers Island has been used for military training, a workhouse, a dumping ground for all manner of the city's "waste" and, currently, a notorious prison complex; and Hart Island has adapted to military training, quarantine for yellow fever patients, a reformatory, a camp for German prisoners of war, a missile installation, rehabilitation for drug addicts and, currently, a civic cemetery for some 850,000 souls unable to pay for their burial. North Brother Island, in full sight of one of the most densely populated areas in the world, has inspired the epithet "the last unknown place in New York City" and has been the site of a huge hospital for quarantining infectious diseases, and a centre for rehabilitating drug-addicted youth; while its relation, South Brother, was the city's first "dump" in the mid-nineteenth century. Ellis Island was for sixty years the site of the nation's largest immigrant inspection station and, when this work demanded that people suspected of carrying infectious diseases be quarantined, the necessary isolation was ensured by building two islands – Hoffman and Swinburne – from landfill. Randalls and Wards Islands have a history of asylums, hospitals and cemeteries; and Rat Island, despite being only one hectare and prone to being awash at storm-driven high spring tides, housed forty patients in a typhoid hospital in the nineteenth century. [3]

An island then, at a purely elemental level, exists within the sphere of certain constructive and destructive geological and biogeographical processes. And if we perceive it to be mysterious, happy, bleak, isolated, beguiling, dangerous, heaven on earth or hell on earth, then that is because we make sense of the island, we give it meaning, by imbuing it with human subjective qualities. The island itself is neither happy nor unhappy, but in our imagination it can make us feel happy, unhappy, or both at the same time.

Our island exists in history, but beyond the knowledge of exploring nations until it enters their orbit as a "new-found-land" at first sighting from the deck of a sailing vessel. A ship may approach the landmass by drifting on light airs from a distant horizon, or by being driven upon storm-lashed rocks; and the island may appear to offer everything that is desired to ensure the good life, or it may be belching fire and brimstone. And so our mariners will have discovered an earthly Garden of Eden or the Gates of Hell. And so, too, it becomes an island of the mind.

The sheer audacity of islands to exist as they do, lost in space, has left writers seemingly lost too, lost for words unless they dig into their bag of hyperbole. So Rockall has been described as "the smallest rock ... in the oceans of the world ... the most isolated speck of rock, surrounded by water, on the surface of the Earth". St Kilda is "the loneliest outrider of Britain", with "the most awesome rockscape of any in Europe"; North Rona is "one of the barest places in Scotland – a low, serrated expanse of rock ringed by never-silent sea"; whilst the Shiants are "an assertion in an ocean of denials, the one positive gesture against an almost overwhelming bleakness". [4]

Outside the British Isles and Ireland, Tristan da Cunha, in the South Atlantic and which was temporarily evacuated in 1963, is "this most isolated inhabited place on earth [where] if the reports were true, it was the happiest and most harmonious community in the whole world – indeed a Utopia of the Sea". On Nam, in the Pacific Marshall Islands, the writer is "numbed by the solitude [giving] cause to wonder if this isn't the loneliest place on earth". And of Easter Island, in the south-eastern Pacific, there is a tradition amongst Chilean authors to describe it as "*la isla mas isla del mundo ... perdido en la imensidad del Pacifico*" ("the most 'enisled' island in the world ... lost in the Pacific's immensity"). Only Raymond Ramsay's description of uninhabited Bouvet (South Atlantic, 2,000 kilometres south-west of Cape Town and 1,600 kilometres from land in any direction) as "the remotest spot on Earth" cannot be challenged! [5]

These islands are the "lost children, wandering stars; [that] call through the headland surf, soft siren voices under the night's edge, the dark of discovery, the dizzying slide of the earth's diminishing curve, calling, calling". [6] And they can assert a mesmeric hold on the imagination to the extent that, for some, islomania may be considered a kind of affliction. Defined by an authoritative Oxford dictionary as a passion or craze for islands, it is surprising that it does not credit Lawrence Durrell with the neologism when, in describing a Greek *paradiso terrestre*, he defines islomania as "a rare but by no means unknown affliction of the spirit [suffered by those] who find islands somehow irresistible. The mere knowledge they are on an island, a little world surrounded by the sea, fills them with an indescribable intoxication. Those born 'islomanes' are the direct descendants of the Atlanteans, and it is towards the lost Atlantis that their subconscious yearns throughout their island life." [7]

David Conover, on his eponymous small island in British Columbia, is more prosaic and practical in describing the source of his own "affliction": "Islomania runs in my blood. I would rather talk about islands than eat; would rather –

and often do – think about islands than sleep. Even a scraggly tree-topped reef, I deem sacred." And, for Conover, size matters. He judges an island to be appropriately small according to the strength of the *feelings* it evokes: "small enough for the eye to know intimately every cove, fir and glen, for one to know those sacred places where, each spring, lady-slippers and camus lilies [sic] abound". [8]

And so islands cast their spell, a spell described variously as: *islomania* – a disease "the least curable and the most enjoyable"; *nesomania* – an intense island obsession; *insulatilia* – being island haunted; and *islophilia* – less an obsession than a deep affection. [9]

Islands in Literature

The historical roots of our fascination with islands are deep. For many hundreds of years there existed an oral tradition, sailors' stories concerning voyages of discovery, a history that was more firmly established when recorded on Egyptian papyrus scrolls in about 2000 BC. Within these records there are elements of what was subsequently to become a well-established genre: a routine voyage, interrupted by the narrator being shipwrecked and cast away on an isolated and previously unknown island. With fragmented geographical knowledge that became more speculative on the periphery of the Mediterranean, and primitive beyond the Pillars of Hercules (Straits of Gibraltar), geography and mythology were finely interwoven to present fictive and quasi-fictive islands scattered liberally across the oceans.

A shallow chronological trawl through this island-studded literature starts with Homer's *The Odyssey* (ca. ninth century BC), where islands provide a context for the journey and a vehicle for the narrative. Plato's ca. fourth century BC account of Atlantis dates its destruction about 11,000 years ago. Euemeros (late fourth to early third century BC) and Iamboulos (ca. third century BC) write about an island utopia and an ideal island commonwealth in the Arabian Sea and the Indian Ocean respectively; and by the time Siculus (first century BC) is writing about Ultima Thule, six days away from what may be the Orkney Islands, primitive incomplete geographical knowledge continues to be embellished by a farrago of myths and, where this did not suffice, pure fantasy.

These shortcomings prompted Lucian of Samosata to write his dubiously titled *A True Story* in the second century AD, prefaced by chastising his contemporaries for their failure to distinguish fact from fiction. It is ironical therefore that his own account of Atlantic islands is run through with imaginative embellishment. Thus, Caseosa is a white island the consistency of cheese; Dionysus is, appropriately, fertile with grapes; and a landing on Dream Island is difficult because it kept retreating from frustrated landing parties. The nuts of their eponymous island are more than five metres long and used as boats; whilst in the particularly unpleasant Empi Archipelago, the mist smells of burnt human flesh and there is the persistent sound of crying and wailing.

Also in the second century the Alexandrian astronomer and geographer Ptolemy describes magnetic islands pulling ships towards them. Not surprisingly, perhaps, given the significance and only partial understanding of ocean currents and tides, the dangers of this magnetising effect of islands becomes a notable theme in recounting voyages of discovery. [10]

The tradition of voyages to exotic places with extraordinary environments and inhabitants continues with the sixth-century travels of Saint Brendan – retold in the ninth-century *Navigatio*, which made a very significant contribution to promoting the Celtic legend of Islands of the Blessed, considered in more detail later in the chapter. But the *Navigatio* also presents us with some particularly exotic islands, and describes an Atlantic replete with giant ants, red-hot pigs, vanishing women, men covered in hair, demon horse races, animals devouring each other, fish tumbling from the sky, and choirs surrounded by flames. [11]

These remarkable accounts should perhaps not surprise us too much, for even 600 years later venturing into the Atlantic – "The Great Green Sea of Gloom" – was considered a sign of insanity. According to the Muslim geographer Idrisi, writing in the first half of the twelfth century, the Atlantic encircles the ultimate bounds of the inhabited Earth, and all beyond it is unknown. "No one has been able to verify anything concerning it, on account of its difficult and perilous navigation, its great obscurity, its profound depth and frequent tempests; through fear of its mighty fishes and its haughty winds." However, he sees fit to assert that "there are many islands in it, some peopled, others uninhabited. There is no mariner who dares to enter its deep waters; or if any have done so, they have merely kept along its coasts, fearful of departing from them." [12] And he goes on to describe the Isle of Female Devils, the Isle of Illusion, the Island of Two Sorcerers, and the Isle of Lamentation, which was fertile but controlled by a terrible dragon. Two pirates lived on the Island of Two Heathen Brothers until they were turned into stone; and the inhabitants of the Island of Kalhan had the bodies of men and the heads of animals. [13]

In the fourteenth century, *The Travels of Sir John Mandeville* was a widely read book of geographic information. It is concerned partly with a pilgrimage to Jerusalem, and reflects the characteristic blend of fact and fiction with which we are now familiar, tempered by Mandeville's Christian zealousness. So on Caffolos in the Pacific, friendship is the prime concern of the people, but the heathen inhabitants of Milk Island are cannibals who delight in fighting, killing and drinking the blood of their victims. On an island called Ghana in the Indian Ocean a wide range of religions is practised, whilst the people on Land of Faith Island live devoutly by the Ten Commandments. [14]

The early sixteenth century witnessed the publication of a number of anonymous island accounts, the most notable published in 1508 and 1538. These include Devil's Island in the Aegean, with its monstrous idols worshipped by giants; the Rock of the Magic Maiden, once inhabited by the daughter of a Greek magician who held visitors prisoner; Monganza Island, home of the very unpleasant giant Famongomadan; and the Island of the Scarlet Tower, off Brittany, where an extraordinary stone tower was discovered by the man who brought the Holy Grail to England. On Brigalaure, butchers make sausages from the ears of sailors, the Luquebaralideaux Isles again feature sausages but these ones are imbued with life, while Pastemolle is surrounded by ovens full of pies.

The humour of these accounts, many of which must surely have been tongue-in-cheek, suggests that writers recognised that they could stretch readers' imaginations by staging narratives on small, remote, previously undiscovered islands. Credibility can be stretched in an island tale far away from mainland scepticism, and such "stretched credibility" is fertile ground for satire. Thus, in the very numerous island descriptions scattered across his books, François Rabelais uses islands – where anything seems possible – as a basis for caustic satires of society, exaggerating particular absurd traits and characteristics for dramatic effect. So his islands have inhabitants deformed, intermarried, descended from hogs, living on wind, who slash their skin to let fat out, who live in a wine press, and who live the life of hypocrite hermits. Here the island is a stage where the inescapable constraints of space corral, compress and magnify some phenomena whilst distorting and intensifying events so they indeed appear larger than life. [15]

Like Rabelais, Jonathan Swift recognised the dramatic potential of a small island as a stage upon which to skewer stereotypical behaviour traits of mainlanders. Although *Gulliver's Travels*, first published in 1726, is an adventure story in which Swift casts his thoughts in a fantasy employing manikins, giants, talking horses and flying islands for effect, it is pre-eminently an attack on English party politics and some of the more dubious achievements of science. It lampoons the pride and complacency which Swift believed afflicted much of the human race in general, and he satirises the vogue for "strange tales" with fictional events treated as fact, the popularity of which had already gained momentum with the publication of *Robinson Crusoe* in 1719.

Lemuel Gulliver's first notable voyage ends in shipwreck on the tiny island of Lilliput, where the tallest man is about fifteen centimetres high. This is a land of empire, where two parties struggle for power, ever-threatened by neighbouring Blefusco. Politicians demonstrate that their minds are suitably agile and fit for purpose by creeping under and leaping over a stick held by the emperor.

On his third voyage, Gulliver is captured by pirates and set adrift in a small boat. He is rescued by the inhabitants of the flying island of Laputa, a strange race of scientists, philosophers and mathematicians wholly absorbed by theoretical issues. Only women, tradesmen and court pages can provide reasonable answers to Gulliver's questions.

From Laputa, Gulliver travels by ship to the island of Glubbdubdrib, inhabited by sorcerers and magicians, and ruled by a governor who is able to call people up from the dead to serve him. From here he sails to the island of Luggnagg, where he finds that some of the inhabitants never die. However, immortality loses its appeal as soon as Gulliver discovers that these people, whilst not actually dying, suffer the ravages of time; so even by the end of the first century they have lost hair, teeth, and a good deal of their mental powers. And there is a long time to go!

In his final voyage, the accident-prone Gulliver is marooned on the Land of the Houyhnhnms. In this island nation, wise, rational and rather dull horses use inferior creatures called Yahoos – like very primitive humans – as beasts of burden. The Houyhnhnm's maxim is to cultivate reason, "opinion" is not understood as a word, and their society is free from crime, avarice and cruelty. [16]

Human nature is indeed resistant to change, and nearly 400 years after Swift was writing, the foibles and absurd activities he laid bare continue to be the focus of modern satire. The point here though is that, like Rabelais writing in the sixteenth century, apparently bizarre people and events are given a degree of credibility because extraordinary things may just be possible on new-found islands. Isolated, small islands, arrived at only by dint of a perilous journey, extend the range of possible realities; and as a literary device they keep the reader balanced, often precariously, between the ordinary and the extraordinary.

By the end of the seventeenth century, the reporting of voyages of discovery was becoming – very slowly – more scientific. At the same time, literature was moving towards the birth of the novel, and Daniel Defoe's *Robinson Crusoe*, published in 1719, revolutionised how the island was perceived (see Chapter Two). With the advent of the novel and the short story, the small remote island found itself the focus of an already heated public imagination. John Bowman has distinguished six prevalent themes in island novels and short stories, themes that concern marvellous islands, supernatural sites, pink-tinted adventure, man alone, nature in the raw, and concentrated reality. [17]

"Marvellous islands" feed the human need to imagine that somewhere there is a more exciting land where more marvellous things take place as a fact of daily life. These islands are usually remote and are stages for utopian undertakings, exotic creatures and paranormal events – a rich mix of fact, fiction and satire. The stories have roots in the oral tradition that preceded the novel and which concerned the exploration of a world circumscribed by limited knowledge, something that was particularly popular at the end of the nineteenth century with the lost continent – Atlantis – utopia theme.

The Island

Island stories based on "supernatural sites" exploit the notion that extreme isolation provokes events that would be considered abnormal on the mainland – ghostly events of supernatural menace. Arthur Conan Doyle's short story *The Fiend of the Cooperage* (1929) is a good example of this.

Related "pink-tinted adventures" take place on islands where people feel released from constraints of civilisation to exercise rather dubious "darker qualities". R.L. Stevenson's *Treasure Island* (1883) is an exemplar, as is H.G. Wells's *The Island of Doctor Moreau* (1896) in which the doctor's island, remote from interfering authority, is turned into a natural human laboratory for exploring the unique quality of the human form and, through rather gruesome experiments supported by harsh social control (which the island's isolation makes possible), attempting to produce "purer forms" of (sub) human life.

Bowman's category of "man alone islands" shares the characteristics of what has also been termed "desert island" literature, in which a shipwrecked traveller is typically cast away on an isolated, uninhabited – hence, "deserted" – island, far from trade routes and any likelihood of imminent rescue. The island is not a waterless waste. It has natural resources that the recluse can utilise, and there is often valuable wreckage fortuitously available. Survival is uncertain, and there is enough physical and mental pain to coerce the solitary islander into contemplating the evils of his previous life that have brought about his solitary demise.

The essence of "nature in the raw" islands is violent environmental conditions that conspire to grossly magnify the discomforts of enislement, trapping islanders as powerless spectators of their fate. Lafcadio Hearn's *Chita: A*

Memory of Last Island (1889) is a fine short story based on the cataclysmic destruction of an island. Tom Neale and Robert Dean Frisbie's autobiographical accounts of surviving a hurricane on Anchorage Island are also fine examples of this genre and are considered in some detail in Chapter Two.

Finally, Bowman's "concentrated reality" stories reflect the fact that the island has been used as a stage on which to explore just about every aspect of the human condition. There is no escape from a small, isolated island, there is no place one can go to recover composure, and events once set in motion cannot easily be delayed, diverted or escaped from.

A cursory listing of island tales where the setting has a radical effect in concentrating reality includes Defoe's *Robinson Crusoe* (about redemption), Wells's *The Island of Doctor Moreau* (perfecting the human form), *The Island* by Victoria Hislop (the treatment of lepers, and how we respond to perceived contagion), *Lord of the Flies* by William Golding (the reality of human nature), *Pincher Martin* also by William Golding (alienation, loss of identity, loss of sanity, purgatory and damnation), *When the Killing's Done* by T.C. Boyle (animal rights in controlling non-endemic species), *Caribou Island* by David Vann (the endgame in a disintegrating relationship), *Snow Falling on Cedars* by David Guterson (war and its aftermath reinforcing racial prejudice), *The Man Who Loved Islands* by D.H. Lawrence (the arrogance of human aspirations), *Greenvoe* by George Mackay Brown (cultural disintegration through militarisation), *Victory* and *An Outcast of the Islands* by Joseph Conrad (exile), and *A Casual Brutality* by Neil Bissoondath (political corruption). Some of this is fine literature, some of it is not. What it shares is the island as a highly effective device to concentrate events that turn mainland reality on its head.

This topsy-turvyness is characteristic of so many island stories, and the island stage is popular with creative writers because a narrative can be presented that so easily leaves us balanced between belief and disbelief. And there are good reasons for this to be found in the existential histories of islands that are so often riddled with ambiguity. Isolation, and the primitive modes through which news of discoveries was disseminated, led – unbeknown to their indigenous populations – to islands being "discovered", "lost" and "rediscovered". For example, archaeological evidence suggests Hawaii was settled from the Marquesas and the Society Islands in the eighth century, and thereafter it appears Hawaiians lived in isolation, lost to the outside world until the arrival of Cook in 1778. Similarly, the Canary Islands were discovered before the Christian era – when they were called the Fortunate Islands – but they were not visited again for thirteen centuries. And the existence of the Azores group too probably suffered periods of uncertainty. It is not difficult to imagine that, during interim periods of being lost and found, the void would be filled by a mix of enduring myths, legends and travellers' tales that would confuse the issue further. [18]

In pre-Christian Irish mysticism, the sea to the west was believed to be the route to the Promised Land or the Other World. Even prior to Saint Brendan's sixth-century voyages, there was an Irish literary genre – the Immrama – characterised by holy men venturing into unknown western seas to feel God's presence and to learn of His wonders.

These stories enjoyed popularity in the Middle Ages equal to that of the legends of King Arthur, and were the prelude for the *Navigatio* of St Brendon's voyages (ca. 545–550 and ca. 551), an account that has been described variously as a factual voyage, a visionary fairy tale, a mythical adventure, a monastic pilgrimage, and delightful fiction. It is probably all of these – the problem lies in trying to distinguish between them. Consequently, it is possible to speculate upon possible landings by St Brendan on St Kilda, the Faroes, Iceland, Newfoundland, the Bahamas and even Florida.

The Immrama was responsible also for claiming numerous imaginary islands in the Atlantic. *The Voyage of Bran*, for example, described 150 western islands, collectively several times larger than Ireland itself. And as the coastal waters became better known, imagination carried the mystery further out over the unknown western sea. "Saint Brendon's Isle" can be traced on successive medieval maps growing larger and travelling ever further west until it becomes part of the New World; at which point an entirely new controversy arises concerning the possibility of a pre-Columbus Irish landing in America.

The tenacity of the Islands of the Blessed in clinging to a vestige of geographical reality is evident in the survival instinct of one particular island. "Brasil" is the Irish Gaelic word for blessed: *breas-ail*. It wanders across medieval charts between the west coast of Ireland and the Azores until it settles in the Gulf of St Lawrence, where it is "found" in 1481 by two Bristol ships that almost certainly made landfall in what is now Newfoundland. Brasil persisted in the minds of British Admiralty chart makers until the second half of the nineteenth century when, in diminished size, it was still evident on charts. [19] Conchur O'Siochain, a serious scholar and promoter of the Celtic tradition, examined archaeological, historical, geological and climatological evidence surrounding Brasil. He concluded that Brasil and the Aran Islands represented fragments of a great land area which subsided under the Atlantic just 2,250 years ago. He saw the legend of islands being seen off the west coast of Aran every seven years as "a 'vision' of what it actually had been ... achieved on some rare occasion when forces and factors in nature, metaphysical and otherwise, of which we have no understanding could reco-ordinate for a fraction of time". [20] In support of his hypothesis he presented accounts of sailors' encounters with Brasil over a long period extending well into the nineteenth century.

Another Celtic scholar, T.J. Westropp, had already described seeing an *apparent* island three times between 1868 and 1872, with hills, trees, towers and smoke, and Samuel Eliot Morison stoked the debate by implying that Westropp was claiming to have seen an *actual* island, something that is clearly not the case. [21]

This is not just scholarly bickering. It would be easy to dismiss O'Siochain's "vision" as having little scientific validity, Westropp's description as ambiguous, and Morison for being rather careless in his interpretation of what has been claimed. But this confusion is symptomatic of a long tradition of uncertainty that hangs over the existential status of Islands of the Blessed, evident even in the twentieth century. Thus in Peig Sayers' autobiographical account of life on Great Blasket Island (south-west Ireland) first published in Irish in 1936, the existence of an island named Brasil is embedded in the consciousness of the islanders. This is reflected in a piece of dialogue between two young men:

"Look Sean! There's Hy-Brasail to the north!"

"You devil you, where?" said Sean for he was a man with curses to burn. He turned on his heal.

"No doubt about it, but it's a lovely view on a summer afternoon," he said. "A person would take his oath that it's some enchanted land."

"Yes, indeed," I said. "I often heard Eibhlis Sheain say that she herself saw Hy-Brasail appearing in that very same place one autumn evening while she was cutting furze on the Brow of Coum." [22]

In the search for Islands of the Blessed, reality and its symbolic representation have merged and parted in a subtle, seemingly symbiotic dance over the centuries, not least because the notion of an earthly paradise, an oceanic paradise, has an almost universal appeal, as if embedded in our psyche: "over there, in the 'island', in that 'paradise', existence unfolded itself outside Time and History; man was happy, free and unconditioned; he did not have to work for his living; the women were young, eternally beautiful, and no 'law' hung heavily over their love .... Geographical 'reality' might give a lie to that paradise landscape ... [but] each saw only the image he had brought with him." [23]

The point here about seeing "only the image he had brought with him" is considered later in more detail concerning an image of paradise that has permeated our perceptions of particularly Pacific and Caribbean islands. Before this however, it is worth asking why, in trying to chart the known world, voyagers, traders, politicians and cartographers have all struggled to distinguish between real and imaginary islands, and have engaged in what has rightly been described as "compulsive island-making ... a troublesome quirk of the human imagination". [24]

Imaginary Islands

Sailors described "position doubtful" islands, and referred to islands that were irregularly located on succeeding chart editions as "flyaway islands". There are three related reasons for the persistence of uncertainty and error: technical issues, like the difficulty in plotting location accurately; the vested interests that some commercial groups had in perpetuating error; and, most importantly, issues directly concerned with human imagination.

Until the beginning of the sixteenth century, navigators seldom needed to fix their precise position. Journeys were relatively short, and ships were only out of sight of land for hours rather than months. Mariners would rely largely upon visual memory to assess the shape and fine detail of the land and the nature of the sea. They used a rudimentary book of sailing directions, a lead line to determine the depth and type of the sea-bottom, a compass, and a sandglass to measure time. But until the middle of the sixteenth century mariners were unaware of magnetic variation; as they sailed west, and as the greater the magnetic variation became, a fixed compass course would send them in an increasing curve south. A course set to Newfoundland would land them on the mid-Atlantic coast of America! And in addition to these limitations in knowledge, until well into the seventeenth century understanding of the general circulation of ocean currents was rudimentary. The direction, strength, extent and seasonal variation of currents was a mystery, but one that could throw a course off by as much as thirty miles in a day – a significant amount when islands were mere specks in the ocean and landfalls could be "tight" even when navigation was good.

Hipparchus (ca. 167–127 BC) had already divided the globe into equally spaced vertical circumferential lines, and Ptolemy (ca. 73–151 AD) spaced the degrees of longitude fifteen degrees apart. Once a navigator reached the latitude he desired – determined by sightings of a celestial body like the sun or Polaris – he set and maintained a course east or west. This was known as "latitude sailing", and the course could be corrected by making celestial observations. But this was recognised even at the time as a frustrating and inefficient way of moving across the globe.

Although the classic journeys of exploration and circumnavigation were remarkable feats of navigation, the determination of longitude continued to be a constant problem even as late as the nineteenth century. Until the development of an accurate chronometer solved this problem, the challenges involved in establishing a ship's position were almost certainly the single most important factor in the very numerous reports of uncharted islands.

In 1530, the Flemish mathematician and cartographer Gemma Frisius successfully devised a method of determining longitude based on time, but it was not until an accurate timepiece – the chronometer – was constructed by John Harrison in 1762 that the theory could be put into practice. The famous No.4 Marine Chronometer was found to be accurate to within five seconds over the course of a journey to the West Indies, but even then it was not until the nineteenth century that production costs were sufficiently reduced to attract ship owners to purchase them readily. This is illustrated by the fact that in 1787, when the first shipment of convicts was sent to Botany Bay, of the eleven ships only one had a chronometer. (It is possible, of course, that the ship owners did not care too much whether or not the convicts reached Botany Bay!) [25]

Within the cartographic process, these errors of reporting were compounded by a lack of opportunity to chart in remote areas that were seldom visited. More importantly, a prudent caution would have been exercised to avoid possible errors of omission – better to include an island despite misgivings and risk inconveniencing naviga-tors than omit one and risk shipwreck. And once even a rather dubious island was charted, consideration of frugal economising would affect any decision to delete it. Altering copper plate was an expensive process. It was only being pragmatic, convenient perhaps, to let it exist, especially if it were of minor significance and in a seldom-frequented area.

The inclusion of bogus islands on charts may also have been a technical device used by cartographers to prove whether competitors were copying their product. And false reporting could be a way for explorers to gain reward, pres-tige and influence. This was well illustrated in Captain John Ross's Arctic expedition of 1829 to 1833 when his nephew Commander James Ross charted three new islands, which he collectively named the Beaufort Group. The islands were individually named after three sons of the Duke of Clarence, who became king in 1830. Upon his return to England, John Ross "reviewed" the expedition's chart book with the King and Captain Beaufort and suggested that, in addition to the three islands, they add six more. If the new king had concerns that these islands did not exist then such misgivings were perhaps allayed by each bogus island being given a royal family name. With a stroke of a pen, the three islands of the Beaufort Group became the nine islands of the Clarence Islands.

Order was restored in 1846 when Sir John Barrow, Second Secretary to the Admiralty, combined diplomacy with hydrographical accuracy by relegating the six additions, which were now charted cryptically as "Innumerable Islands". [26] (Sir William Pepys had an island in the Atlantic named after him during his time at the Admiralty. Unfortunately, for him at least, it was fictitious.)

Motives derived from a consideration of commerce encouraged fraud, as perpetuating stories of islands rich in economic potential served to encourage patrons and subdue creditors whose speculative interest and financial assistance might otherwise falter. It is reasonable to suppose that fishermen too, especially sealers who depended on relatively remote coastal regions for their industry, may have encouraged the proliferation of false island reports to handicap competitors. It is quite likely, for example, that Bristol merchants often used the search for islands like Antilia and the Seven Cities as an excuse to extend the domain of cod fishing. Likewise, the "promoters" of these same islands recommended them in terms of their strategic benefit as way stations en route to the "Indies".

Fixed ideas, and a reluctance to change them in the face of strong confounding evidence, were a major barrier to discovering even the bare outline of global geography. Erroneous world views can start as a radical fashion, only to become established dogma blocking meaningful progress.

These *idées fixes* are numerous and diverse; they can even appear primitive and eccentric to the modern eye, but they had a profound effect on the historical process of making sense of the world.

Ptolemy exercised a pre-eminent influence on fifteenth-century geography, but his picture of the world included a substantial error of omission in his estimation of the distance from Europe to China. This error was then perpetuated for centuries, even by influential cosmographers like Paolo Toscanelli, who advised King Alfonso of Portugal from 1474, and whose ideas on Asiatic locations were based on Marco Polo's notes (which were nearly 200 years old). These factors encouraged early navigators like Christopher Columbus and John Cabot to look optimistically towards a short Atlantic crossing, and later the Portuguese Ferdinand Magellan, who organised expeditions for the Spanish, to anticipate an Asian continent located close to new-found America, maybe along a chain of still-to-be-discovered Asiatic islands. For Magellan, this would place it conveniently within the Spanish sphere of influence as established by the Treaty of Tordesillas of 1494.

Disappointment in finding America instead of a route to the Indies inspired the promotion of another *idée fixe*: the Straits of Anian were rumoured to be a North-West Passage, the key that would open the route to China. This too stimulated numerous real and apocryphal voyages, and indeed it was whilst James Cook was exploring for a sea link between the Pacific and the Atlantic that he "rediscovered" Hawaii for the Western world during his third voyage in 1778. [27]

I have suggested that the thinking and behaviour of

people who sent ships out from European ports in search of "new lands", and of the people who navigated these vessels, was extremely conjectural. Indeed, the allure of the fantastic could take precedent over the tedium of incremental exploration based on established fact. Within this context the search for islands was part of an overall pattern within which many primary discoveries were made fortuitously, by chance, by men chasing chimeras as the imagination took precedent over what were often only loosely established geographical facts.

Indeed, the promise of trade routes and the attraction of gold were issues which stirred the public imagination, something that resourceful literary publishers were quick to exploit. The 1838 American scientific Wilkes Expedition to the Pacific and the Antarctic had among its practical aims the location of stations for the whaling industry. But it represented too something of a fashion for searching for islands, inspired by the at least partly apocryphal voyages of self-promoters like Benjamin Morrell on the Arctic fringes between 1812 and 1831. Edgar Allan Poe picked up on this to write *The Narrative of Arthur Gordon Pym of Nantucket*, first published in 1838, a bizarre tale of mutiny, famine, cannibalism and primitive people amongst previously "undiscovered" or "lost" islands. It was popularly received, and the theme of journeys to rediscover islands has prevailed in modern literature. For example, Geoffrey Jenkins' *A Grue of Ice* (1962) is based on Bouvet Island, which we have touched upon already as "the remotest spot on Earth", located as it is in the South Atlantic's nether regions. In 1825, Captain George Norris landed on Bouvet and spotted a second island seventy-two kilometres to the north-north-east. He called it Thompson Island. It was reported too by Captain Fuller in 1893, but it has not been seen since. The first accurate map of Bouvet was only produced in 1985 – 247 years after it was discovered – when a Norwegian expedition had sufficiently clear weather to allow it to be photographed from the air. Thompson Island appears on maps as late as 1943, and there are suggestions that it disappeared in an undetected volcanic eruption. Subsequent surveys indicate that its supposed position lies in nearly 2,500 metres of water.

One reason why Captain Fuller's chance sighting had not been repeated – and the reason why Thompson Island did not exist – is perhaps attributable to mirages which occur during certain rather particular sea and sky weather conditions. In coastal waters the phenomenon of the *Fata Morgana* is named after the fairy sister of King Arthur, Morgan le Fay, whose magic created visions of island harbours and cities that lured seamen to their death. It was first described in the Straits of Messina in 1773. This mirage is attributed to portions of coastline being magnified, distorted and optically transferred, and is something that can occur during unseasonably calm and misty weather. [28] That these "mirages" gave rise to concern amongst mariners is demonstrated in two passages from Henry Stommel's excellent *Lost Islands: The Story of Islands That Have Vanished from Nautical Charts* (1984):

> "A friend of mine, Captain Allen Jorgensen, recalled how as a third mate in his youth he spied at night what appeared to be a high island directly ahead of his ship, in a South Atlantic position where no such island was charted.

After convincing himself that it was no illusion, he woke the captain in some alarm, and the old man turned over in his bunk with the admonition to stay on course. Back on the bridge, the mate grimly steered for the apparent island. Great black cliffs reared up, and breakers appeared at their foot. And then miraculously the ship passed through them; it was a night-time rain-squall.

"And there is the instance described by Captain John Byron during his circumnavigation of 1764–66, on Monday, 12 November 1764. 'It thundered and light-ninged [sic] very much and looked black almost around the horizon. I was then walking around the quarter deck when all the people upon the forecastle called out at once: land right ahead. I looked under the foresail and upon the lee bow, and saw it to all appearance as plain as ever I saw land in my life. It made at first like an island with two very scraggy hummocks upon it, but looking to leeward we saw the land joining it running along way to the southwest. We were then steering to the southwest'. Byron sent officers to the masthead to look out upon the weather beam, and they saw land a great way to windward. He sounded and found the bottom at fifty-two fathoms and thought he might be embayed. He therefore made sail and sailed ESE, and there was no change in the appearance of the land. The hills looked as blue as ever, and some of the crew saw the sea break upon the sandy beaches. After about an hour the land suddenly disappeared, to their great astonishment." [29]

Nineteenth-century charts and atlases are reckoned to contain up to 200 islands that are now known not to exist. So Pitcairn Island – which itself was charted about 125 miles from its true position when the mutineers from HMS *Bounty* landed in 1789 – had the ghost islands Encarnation, Michel and San Pablo within 200 miles, as plotted on Admiralty Charts until the mid-nineteenth century.[30]

But navigators had to make do with what they had, and so tended to assume chart information was more or less accurate in locating islands – an assumption that led to fruitless searches even in the later part of the nineteenth century. For example, in the Pacific as late as the 1870s ships used Easter Island in their search for Pilgrim Island and Wahou, which were non-existent yet extant on contemporary charts. The 1875 Revision of the Admiralty Pacific Chart deleted 123 islands and vigias (rocks, reefs or other hazards), including three real ones! Non-existent Ganges was still considered a "doubtful" danger to navigation by the International Hydrological Bureau in 1932.

That islands continued to be *terra incognita* in the mind well into the twentieth century is evidenced by Gilbert Grosvenor, who in the 1921 presidential address to the National Geographic Society suggested that there may still be some Pacific island discoveries to be made. Certainly before the advent of satellite surveying technology, there was a period of 150 years or more when the demise of sailing trade routes and of island-based industries like sealing and whaling resulted in oceanic regions in high latitudes and in the Pacific being less well-known. But that is history.

In the Atlantic, mobile and migratory Islands of the Blessed became "doubtful" islands that suffered on succeeding charts by being reduced to rocks, banks and eventually sea mounts. But this has not been a fast or even process. The 1981 edition of the National Geographical Society Map of the World continued to mark non-existent Matador Island and Atlantic Island. In the South Pacific–Atlantic fringe area, Royal Company Island was only deleted from charts in 1908, the Nimrod Group in 1936, and Emerald Island in 1954. The latter was marked on a Soviet State Geodetic Institute map as late as 1974, and on the 1985 *Times Atlas of the World*. [31]

Excepting newly formed islets and those revealed by melting ice, time is being called on the discovery of anything new. The pendulum has swung decisively toward the undiscovery and elimination of islands. Thus, even as recently as 1980 an American research vessel reported an uncharted island in the South Sandwich Group in the South Atlantic. A British survey ship was dispatched to confirm its existence and to annex it if appropriate. They failed to find it.

And more remarkable, perhaps, is the experience of a group of Australian scientists who in November 2012 sailed in very deep water through Sandy Island, the existence of which was accepted by Google Earth, world maps, marine charts and scientific publications. The prestigious *Times Atlas of the World* shows the island – in 4,620 feet of water, as it turns out – but with the name Sable Island. The director of charting services for the Australian Hydrographic Service is reported to have said that some map makers added non-existent streets to keep tabs on people stealing their data, but admitted that this was not standard practice with nautical charts. (Although, as we have already seen, in the more distant past chart makers perpetuated "false geography" as a means of checking plagiarism by competitors on later editions.) It is perhaps noteworthy, too, that were Sandy Island to exist, it would be in French territorial waters, and that with their interests in mind they have presumably explored the area carefully. It does not appear on French charts. [32]

That these pragmatic acts of elimination have not been achieved without a tinge of regret is illustrated by the ocean-ographer Henry Stommel in reporting with unconcealed sorrow that "the main task has been one of extinguishing, one by one, little points of land, some of which, we cannot help thinking, ought to have existed. ...The joy of achieve-ment is alloyed with a melancholy for realms of fantasy and romance forever dismissed." [33]

Certainly there has been regretful "extinguishing", but there is still the chance for explorer's dreams to come true in the discovery of some "little points of land". They would be cheered by the recent discoveries (in 1976) of Colvocor-esses Reef in the Chagos Group (Indian Ocean), and of Landsat Island (just twenty-five metres by forty-five metres and six metres high) in Newfoundland waters.

But perhaps it is appropriate to leave the final words on the charting of islands with the captain of a vessel in Mari-anne Wiggins's dystopian novel set in the Andaman Islands, in the Indian Ocean, describing a survey "made in 1888 by a drunkard with one eye and quill dripped in invisible ink ... [with islands] like seeds broadcast from a single pod, each one the doppelganger of the others. ... One deepwater channel between un-named islands on the survey was now a solid field of coral and ...two sister atolls ...had submerged entirely into shoals. Earthquakes raked the area two or three times every year, heaving brand new atolls and lagoon islands from what were previously shallow fringing reefs." [34]

Chapter Two

Crusoe: Castaways, Maroons and Beachcombers

The Island

In 1776, eight people were rescued from a tiny, treeless scrap of coral and sand in the Indian Ocean. Seven adult women and one baby boy were all that remained of a group shipwrecked and then marooned in 1761 when a French vessel carrying an illegal cargo of Malagasy slaves went down on this uncharted isle. Their story varies in small detail, but the key elements are put together from the ship's log found in marine archives at Lorient in Brittany, and from "interviews" with the rescuees carried out by the French colonial administration on Île de France (now Mauritius).

At least twenty sailors drowned, together with seventy slaves, whose chances of survival had been made even more perilous because hatches were not opened to enable escape. Of the eighty-eight slaves reaching shore, about a third died in the first few weeks because the sailors kept meagre water supplies to themselves; and more died because they hailed from the Central Highlands of Madagascar and had no experience of living in any marine environment, let alone one as harsh as this.

A boat was put together from the wreckage, and after six months the surviving 120 sailors and two "gentlemen" packed themselves aboard. The ship's human cargo was left behind. When they reached Île de France, about 375 miles away, the colonial administration was involved in the Seven Years' War with Britain, it was anticipating an attack from India, and it had banned the import of slaves as they were simply more mouths to feed in the event of a siege. The governor refused to risk the loss of another ship for a group of unwanted and illicit slaves.

Eighteen of the maroons left behind on the island attempted escape by building a small raft or boat. There is no record of them ever being seen again. The remainder kept the same fire going for fifteen years with scraps of material from the wreck. With most of the ship's timbers already re-used, they were forced to build shelters from coral and by making cement-like blocks from compacted sand or volcanic rocks. For drinking water they used a well that the sailors had dug, they built a communal oven, and they repaired copper pots, some as many as eight times, using hand-made rivets. And they survived on a diet of turtles, seabirds and shellfish.

The maroons built a lookout on the highest point, and in 1776 a French warship spied them and risked a rescue through the reef with a small boat. This too was wrecked; but with the resourcefulness and determination that runs through this story, yet another floating contrivance was cobbled together. Eight Malagasy and their French rescuers sailed to Île de France from whence a vessel was despatched to pick up the remaining castaways. The new governor not only organised the rescue but insisted that the Malagasy were not in fact slaves because they had been purchased illegally. All were now free, and the governor even adopted a family of three. The island, which carried the name of Sand or Slave Island for a while, was eventually charted as Tromelin Island in honour of the captain of the French warship that first sighted the maroons. [1]

The events recounted here in bare bones go to the core of human existence – the need to find shelter, food and water, and to sustain life in a harsh and perilous environment. Stories like these, of "desperate journeys and abandoned souls", hold vital elements that grab the human imagination: a harsh, remote, inescapable place; a small group of vulnerable people whose survival demands resourceful-ness and resilience; a corrupt organisation prepared to do nothing to retrieve events; a dangerous and heroic rescue, and a happy ending for the few fortunate enough to reach the end. It is not surprising then that tales of people being cast away on dangerous shores was fertile ground for sailors' yarns, folk songs, storytelling and imaginative literature. [2]

Crusoe – the Archetypal Castaway

The literature of castaways has a distinguished lineage, with Saint Paul on Malta, Odysseus on Scheria (sometimes named Phaeacia), Prospero on Bermuda and Gulliver on Lilliput, but it was the publication of *Robinson Crusoe* in 1719 that cemented the castaway's place in history. It was an immediate success, so much so that it prompted Defoe to pen a number of hastily written sequels, and encouraged a number of shrewd imitators to develop their own castaway heroes. By the beginning of the twentieth century the "Robinsonade" was an established genre in European literature, with Crusoe depicted in more than 700 editions, translations and imita-tions. The Inuit of Greenland have their own translation. In the eighteenth century, Richard Sheridan directed a popular Crusoe pantomime; in the nineteenth, Jacques Offenbach wrote the music for an opera; and in the twentieth century, Luis Buñuel made one of several films that feature this "greatest best-seller in modern times". [3]

In researching his book and its successors, Defoe was influenced considerably by the experiences of castaways and maroons in an age when merchant, military and priva-teer ships were not restricted to well-established routes, and small islands provided crucial fresh water, food and fuel. To be marooned was a terrible fate, used by privateers, pirates and buccaneers for the most heinous offences. Put ashore alone and with very few possessions, unless the small island was verdant – and many were not – it usually meant death by slow starvation if one did not die of thirst first. Even if the maroon was resourceful, he could still easily succumb to loneliness and depression, or to death, slavery or imprisonment at the hands of an enemy. Thus, for example, a Dutch ship's officer was set ashore on Ascension Island in 1725 as punishment for the crime of sodomy. Nine months later, only his tent, diary and some belongings were found by British mariners.

Defoe would have been familiar with the voyages and writing of William Dampier, including *A New Voyage Round the World*, published in 1697, which began a fashion for "discovery literature". This became a popular genre that was prosaic in style whilst presenting the nearest thing to a scientific account of voyages of discovery. Dampier was intimately aware of the realities of maritime survival, having lived through two perilous open boat voyages and two maroonings. In 1701, as commander of HMS *Roebuck*, his ship was sunk in Clarence Bay – the main anchorage for Ascension Island – and he and sixty men survived for two months until rescued. He was present when Alexander Selkirk – commonly cast as Defoe's key inspiration for *Robinson Crusoe* – was marooned on Más a Tierra in the Pacific in 1704 and, remarkably coincidentally, Dampier was with the privateer Woods Rogers when he was picked up in 1709. Selkirk's rescue and return to Britain – he subsequently died at sea in 1721 – were recounted in articles by Woods Rogers and by Richard Steele, who published a piece in *The Englishman* in 1713. This captured the public's imagination and there were plenty of other accounts of maroons for an inventive author to work from.

Indeed, on Más a Tierra alone Selkirk had predecessors. In 1681, a landing party under Captain James Watling had to beat a hasty retreat when Spanish ships appeared. In the rush a Miskito man named William was stranded with only a gun with some powder and shot, and a knife. From these he fashioned a saw, and some harpoons and fish hooks, and managed to survive and elude capture by Spaniards until he was rescued in 1684. And a couple of years later, five English pirates and four natives were put ashore at their own request and survived to be rescued three years later.

There are many similarities between the experiences of Alexander Selkirk the voluntary maroon (I use the term cautiously, because there is evidence that he may have changed his mind as he was deposited on the beach) and Robinson Crusoe the castaway. Both were extremely fortunate in having an island home gentle of climate and bountiful in providing the natural resources for human existence. "Venomous and savage creatures" lived in the minds of the two men, but not on their islands. Both came ashore well-provisioned – even Selkirk was allowed all his possessions, which would have included his bedding, a firearm, his knife, hatchet, kettle, flint and steel, and some books including a Bible. Both spent their early weeks in despair, appalled at the prospect of complete isolation from their fellow men. And even for Selkirk, who was reported to be so at ease with himself and his island when he was rescued, it took about eighteen months before he was reconciled to his condition. And both were remarkably resourceful in transforming their island for their survival and relative comfort.

For Daniel Defoe the island is a stage where the narrator, alone and with plenty of time on his hands, can reflect upon the wicked life he has led before God casts him upon a lonely shore. But even so, God is merciful, and Crusoe's island is bountiful; he is equipped with numerous bibles and he has a surfeit of time to study them – twenty-seven years, as it turns out. As with many of Defoe's imitators then, this island is a temporary refuge where the castaway is relatively comfortable but seldom free from anxiety and fear of wild animals and "heathen savages". [4]

E.M. Forster's explanation for the popularity of *Robinson Crusoe* as adult literature – that it comforts the English to feel that a life of adventure could be led by a man duller than themselves – is fair insofar as Crusoe is marked by a disturbing lack of curiosity about his island. And even where there are glimmers of interest in getting to know his small world, this impulse is kept in check by constant anxiety and a fear of nature and the animals and savages that dwell there. [5]

The verdant, potentially productive quality of his "Island of Despair" is something taken for granted, a setting for his fearful defence-building activities that are described in terms appropriately purgatorial. In seeking diligently to fulfil God's will, his overriding intention is to win salvation, which is synonymous with escape from his island. And in keeping with the civilising instincts of eighteenth-century Christianity, man and nature wage war until the wilderness is rendered productive and its savages converted to godly ways.

Crusoe finds that his basic needs can be relatively easily met, so there is little stimulus to explore further than he has to. After ten months, the other side of the island – about a day's walk – remains unvisited, and only then does he decide "to make a more perfect Discovery of the Island". But there is nothing to indicate that he acted upon this decision. He takes six years to attempt a circumnavigation by boat, and the event is fraught with so much anxiety and dread that he seldom makes use of the boat again.

He invariably sleeps in a tree surrounded by stakes when he is away from his ingeniously fortified bases, and the discovery of a human footprint precipitates two years of hard labour in developing additional hidden defences. His concern for personal safety – which sometimes belies his rejuvenated faith in God the protector – remains with him for eighteen years, and the reward for this care-worn vigilance is the discovery of the human remains of cannibalistic ceremony. Unsurprisingly, this reinforces his feelings of dread and intimations of impending catastrophe: "I slept unquiet, dream'd always frightful Dreams ... [in] the life of Anxiety, Fear and Cares, which I had liv'd ever since I had seen the Foot of Print".

The he rescues a captive from cannibals – an event expressed in the social stereotypes current in early eighteenth-century England – and "the savage", whom he names Friday, immediately recognises his superior saviour and adopts an appropriately deferential attitude as servant. In this, Friday is perceived as an animal, a creature waiting to receive humanity from Crusoe, who is the sole possessor of knowledge and wisdom. His best hope is to be civilised by becoming a good servant. And so the old, established English society is recreated. Crusoe's civilising zeal, held together by blind conviction, is now reinforced, and together they strive to transform the wild jungle, to build ever more extensive fortifications and more complicated buildings, to develop a productive garden and to establish a large food store. By these activities the castaway strives to reproduce the comforts and security of his lost world.

Crusoe's initial requirements (for what he anticipates optimistically would be but a brief sojourn on his island) are listed as health and fresh water, shelter from the heat of the sun, security from ravenous creatures whether man or beast, and a view of the sea to facilitate rescue. But eleven visits to the wrecked ship – which conveniently lies at ease for thirteen days – provide him with a rich diversity of goods. Even Crusoe admits that there is more than he knows what to do with, including a fine arsenal of weapons and a comprehensive carpenter's tool chest with which to build anew the world he had left behind.

His first response to his new world is one of shock as he surveys his changed circumstances on "this Horrid island ... [in] this dreadful Condition ... this dismal unfortunate Island which I call the *Island of Despair*, all the rest of the Ship's Company being drown'd, and my self almost dead".

But even Crusoe has to be thankful for small mercies, as he admits, "I could hardly have nam'd a Place in the unhabitable Part of the World where I could have been cast more to my Advantage"; and his attempt to circumnavigate the island leaves him relieved to "look back upon my desolate solitary Island, as the most pleasant Place in the World". A "fortunate" isle upon which to be castaway then, added to which Crusoe is able to make a virtue out of his solitude in trusting that divine inspiration lay behind his shipwreck. It may seem strange that castaway stories, many of which are far more tragic than this one in their litany of terror and sorrow, were held up as evidence of God's love for humankind, but Crusoe recognises that God has indeed been merciful, and the castaway has plenty of time to seek atonement for his "wicked, cursed, abominable" mainland life.

It is apparent that Crusoe's experiences on his Island of Despair are circumscribed by his need to impose a regime that reflects the familiar order of his mainland life. The savage Friday is inducted into this process through his ready acceptance of his master's values. When a ship finally arrives, in the fracas that ensues amongst the conflicting elements of the crew, Crusoe exerts his command over them, too. He establishes a miniature state in which he can say confidently "my people were perfectly subjected; I was Lord and Lawgiver".

Salvation arrives with Crusoe's rescue, and he departs his island with no regrets. But memory plays strange tricks, and by the time he is returned safely to the mainland he is reflecting upon the carefree "silent state of Life in the Island", a recollection at odds with a reality that was more grinding toil and anxiety-ridden than care-free. But perhaps in touching upon island life as the good life, Defoe is merely preparing his readers for better things to come, as there are hints too of more adventures to follow. *The Further Adventures of Robinson Crusoe* and *Serious Reflections During the Life & Surprising Adventures of Robinson Crusoe, With His Vision of the Angelic World* were published in 1719 and 1720.

Crusoe's Legacy

Modern critics may have accused *Robinson Crusoe* of excessive dullness, but in its time the island castaway theme fired
a public imagination already primed by the more prosaic
accounts of voyages of discovery. Ambrose Evans's *The
Adventures and Surprizing Deliverances, of James Dubourdieu,
and His Wife* appeared only a few days after Defoe's second
volume, and Evans added *The Adventures of Alexander Vend-
church*, also in 1719. Other imitations, which often added
bizarre supernatural adventures and new species of beings,
followed: Robert Paltock's *The Life and Adventures of Peter
Wilkin* (1750), Ralph Morris's (pseudonym) *The Life and
Astonishing Adventures of John Daniel* (1751), William Bingfield's (pseudonym) *The Travels and Adventures of William
Bingfield, Esq* (1753), and John Howell's *The Life and Adventures of Alexander Selkirk* (1829). [6]

Inevitably, the Robinsonade became formulaic. The
well-known examples of R.M. Ballantyne (*The Coral Island*,
first published in 1857), Frederick Marryat (*Masterman
Ready*, 1898) and Johann Wyss (*The Swiss Family Robinson*,
1869) encompass an environment captured well by Walter
de la Mare: "an island let it be, say, three or four hundred
to a thousand miles or so from the nearest habitations of
humanity and well out of the usual sea-trade routes, pref-
erably uncharted, fairly commodious, say thirteen miles
by four, of a climate whose extremes are not of a pitiless
severity, an island which Nature's bounty has endowed with
shade, fresh water, shelter and food fit for human consumption. And there – our recluse." [7]

De la Mare uses the term "desert island" to describe his
recluse's involuntary home, and this reflects two particular

47

qualities. The island is unpopulated or depopulated, a "deserted" place which has been relatively unaffected by the mainland. Secondly, in the aftermath of the success of *Robinson Crusoe*, to be on a desert island takes on a degree of exoticism. Thus, by the time J.F. Bowman's *The Island Home, or, The Young Castaways* is published in 1853, he is able to parody the genre and unwittingly predict the extravagantly bountiful island that is the setting for *The Swiss Family Robinson*: "Did you ever hear of a desert island that wasn't a lovely spot ... [It] should combine the richest productions of temperature, torrid and frigid zones – a choice selection of the fruits, flowers, vegetables, and animals of Europe, Asia and Africa." [8]

Michel Tournier's *Friday, or, The Other Island* [9] is less an imitation of Defoe than the remarkable emergence of a Crusoe who, during the course of the narrative, is transformed from an image of Defoe's outsider to somebody whose "conversion to the sun had been completed in tranquil happiness". If Robinson Crusoe built fortifications to protect himself from the malign influences of nature, and in the process intended to edify the reader by promoting the virtues of a God-fearing life in civilised eighteenth-century England, then *Friday* is a story of transformation achieved through a profound relationship with the wild environment, a relationship which eventually constitutes the very meaning of his life. If God is a presence, it is not in the form of a Christian deity, but in the pantheistic sense that Godliness is in everything emanating from Nature. *Friday* is thus a radical departure from most Robinsonades that are cast in the mould of the original Christian parable.

Unlike Defoe, who sites his island about sixty kilometres

from the Orinoco estuary – at variance with the experience of his inspiration, Alexander Selkirk – Tournier locates his island in consideration of Selkirk, on "Speranza" near the outer reefs of Más a Tierra (now Robinson Crusoe Island) – a place "unknown to map-makers, lying somewhere between the Juan Fernando Archipelago to the west, and the Chilean coast to the east". Like Defoe's castaway, Tournier's Robinson christens his island in immediate response to the catastrophe. For him it is the Island of Desolation, like Crusoe's Island of Despair, where "naked and alone in that apocalyptic landscape, his only companions [were] two corpses rotting in a wrecked ship".

After the shipwreck Robinson is reluctant to undertake any work that might suggest he is establishing roots. He seldom leaves the beach from which he anticipates imminent rescue by a passing ship that would transport him back to the familiar mainland. It is only with reluctance that he eventually explores the wreck, and thereafter his efforts are directed towards building a sailing boat he names "Escape". But as with Crusoe, who also tries his hand as a shipwright, the boat is too heavy to launch. Sailing away seems impossible, and the lack of any passing ship convinces Robinson that this island is remote from shipping routes.

Recognising his predicament in the all-consuming vastness of the ocean overwhelms him. With no effective antidote to aching loneliness, the terror of exile reduces him to an animal state of wallowing in the island's mire. But he soon realises that melancholic self-obsession could have fatal consequences, and this shocks him into purposefully obsessive activity in much the same way as Defoe's Crusoe.

Robinson embarks upon the next stage of his island life by

renaming the island Speranza, the Island of Hope. To escape the chaotic natural order of his island he is determined to impose a civilised order with which he is more accustomed. And so he builds a Conservatory of Weights and Measures and a water clock that allows island time to be "tamed by the strength and resolution of a single man". Like Crusoe, he applies relentless discipline to ensure the construction of substantial fortifications to guard against savages, and strives for the cultivation of enough grain and foodstuffs to feed a village for several years. "Progress" means transforming the island through ritual, routine and compulsive activity by which he seeks security from the horrors of an unknown world and from the clamorous demons within, which give him no peace even in his solitude.

Despite these displacement activities, Robinson does "regress" to the extent that he is not able to suppress completely moments of doubt, an incipient sense of futility that the fruits of his labour are barren if they are not shared. But if his rigidly structured life focused upon defence and cultivation feels hollow, he recognises too "a new man [who] seemed to be coming to life within him, wholly alien to the practical administrator". This "stranger" is feeling something like tenderness for *his* island, his home. He has intimations too that the natural world of Speranza is fulfilling a maternal role in providing for his needs, something that needs nurturing through a relationship of co-dependence rather than domination.

Robinson's growing sense of wonderment is fleeting and elusive, and he continues to oscillate between old habits and dawning possibilities. It is inevitable then that the arrival of Friday – who accepts the tutelage of Robinson, but who is considerably less tractable than Crusoe's "savage" – complicates matters further. Friday is wholly unresponsive to his biblical proselytising and demonstrates little commitment to transforming the island to a productive unit. Instead, he seems profoundly content to experience the island in what appears to Robinson to be mere play, in exercising a child-like curiosity, in daydreaming and apparently directionless wanderings. To the frustrated Robinson "with every appearance of goodwill, he was proving himself utterly unreceptive to principles of order and organisation, planning and husbandry". Worse still, Friday's apparent lack of zeal challenges Robinson's painfully gained victory over the "natural savagery" of the island.

This confusion is brought to a head cataclysmically when Friday retreats into the storage cave for an illicit smoke from his master's dwindling supply of tobacco. Realising his subterfuge is discovered, he tosses the remains of his smoking into the back of the cave. Forty barrels of gunpowder explode, instantaneously. All the equipment, all the hard-gained food supplies, and all the constructions they have toiled over – all is destroyed in a moment. And with the explosion Robinson's resolve to continue with "the project" goes up in smoke, too. Shaken into the realisation of just how much his civilising labours were oppressing him – almost as much as they were Friday – Robinson understands that now might be the time for his servant to show him the way to something else:

"A new Robinson was sloughing off the skin of the old, prepared to accept the decay of his cultivated island, at the heels of an unthinking guide, to enter upon an unknown road." In the company of Friday, guided and instructed by

Friday, he engages in sports and pastimes that he would formerly have considered beneath his dignity. On a tropical island, where Robinson's basic needs were easily provided for, he recognises that he could achieve a sense of harmony which transcended a life dominated by rigorous productive labour. And so he imitates his companion, who never seems to work in the labouring sense, who is unconcerned with past or future, and who lives wholly in the moment.

Friday would lie for days in a hammock bringing down birds with a blowpipe, and in the evening he would drop a bag of birds at Robinson's feet "with an offhand gesture which might have been that of a faithful retriever or, on the other hand, of a master so sure of his authority that he did not need to give orders". And it is through his dawning appreciation of Friday's lifeworld that Robinson discovers that the isolation and deprivation which had once led him to wallow in the mire of desperation can be turned around.

That Robinson's metaphysical rebirth is made possible by the isolation of his small island is revealed dramatically when, twenty-eight years after his shipwreck, Speranza is visited by a Liverpool schooner. After so long he is eager to be among his fellow men, a human presence without which he was almost going mad soon after being cast ashore. But his excitement at examining all the paraphernalia of their arrival is muted by a sense of discomfort that is more than just shyness after long isolation. For the ship's crew, shipwreck and being cast away on a strange shore are just part of the business of being at sea. They are only marginally interested in the detail, preferring instead to discuss events of the recent war with America and profits from trade. Robinson shows them what the island has to offer in game, fresh fruits and vegetables as a protection against life-threatening scurvy, and the crew does what seems natural – by seeing to their immediate needs, and then taking more for an unknown future. Robinson knows that in former times he had been as they were, "driven by the same motives of greed, arrogance and violence". He imagines being asked what he now lived for, on a tiny isolated scrap of land, and responding, "by pointing with one hand to the shores of Speranza and with the other to the sun. After a moment of stupefaction Hunter [captain of the sailing vessel] would have burst out laughing, the laughter of folly in the face of wisdom: for what was the sun to him but a gigantic flame?"

What other response might one expect? The captain and crew are outsiders, after all, their imaginations brutalised by years at sea. But for Robinson their attitudes serve to bring into sharp focus his own transformation – the feelings of attachment, stability and unity – and the dawning recognition that his rebirth would be but a transient episode if he rejoined the mainland world.

Robinson decides to stay, now a voluntary maroon rather than a castaway. And Friday departs, secretly, with the schooner. As Robinson wrestles with the shock of his separation from Friday, he discovers the boy who worked for the ship's cook, and who had hidden ashore to avoid further brutal treatment at sea. "He took the boy by the hand, and together they began to climb the slope leading to the sharp peak of rock which was the highest peak of the island. ... To the north a gleam of white was speeding towards the horizon, and Robinson pointed towards it: 'Look well at it,' he said. 'A ship within sight of Speranza. It is something you may never see again'."

The Island

Staying on his island contrary to all expectations is something Alexander Selkirk may have wished for in retrospect. For his article in *The Englishman* in 1713, Richard Steele spent some time talking with Selkirk and describes his life as "exquisitely pleasant…his nights were untroubled and his days joyous, from the practice of temperance and exercise". Later in his home town of Largo, Scotland, we see him returning time and again to a hillside cave where he can gaze out to sea in a melancholy and nostalgic attempt to recreate his Más a Tierra life. [10]

Voluntary Maroons

For Defoe's masterpiece, being cast away on a small remote island, even one endowed with provisions from the doomed ship, is perceived to be a dehumanising experience. Healthy intercourse with one's fellow men is replaced by pathological anxiety over the absence of friends left behind and the hostility of enemies unseen. The island is thus a reduced form of what the world has to offer, a kind of outdoor prison.

However, some real-life castaway experiences – especially those of "voluntary maroons" – complement Tournier's account of how the natural limitations of an island environment may stimulate an intimate and profound sense of attachment to place and a meaningful existence, even without other people. Tom Neale is one such example. He spent fifteen years living alone on Anchorage Island in the northern Cook Group, Polynesia, and we will return to him later in the chapter.

Voluntary maroons are part of a beachcomber tradition that appears prominently in the history of European influence in the Pacific. The word "beachcomber" is really a generic term for just about any European who "jumped ship" and "went native", and often wielded some influence in island politics. Gonçalo de Vigo is, perhaps, the first example as he deserted from Ferdinand Magellan's ship, *Trinidad*, in 1522 on an island in the Pacific Mariana Group, from whence he was taken to Guam where he lived for four years. But the term was not used until much later, Herman Melville probably using it first when he jumped ship in the Marquesas in 1842, and wrote about it five years later in his book *Omoo: A Narrative of Adventures in the South Seas*.

The heyday of the beachcomber was quite short, from the late eighteenth to the early nineteenth century, when there was a steady flow of deserters from British and American whaling ships in the Pacific. But they included people from many walks of life, deserters from all manner of vessels, stowaways, escaped convicts and adventurers from aristocratic backgrounds. Many were escaping from brutalising environments, but they often integrated into indigenous communities well, better than the missionaries, traders and government officials with whom they were contemporaneous. In Polynesian islands, in particular, communities were often receptive to assimilating outsiders, aided by myths concerning the return of gods or ancestral spirits that could be perceived as coming true by the arrival of an exotic stranger. But beachcombers were often men with artisan skills who could fashion metal tools, make trade items into useful goods, and repair, maintain and direct the use of firearms.

They could act as go-betweens when sailing vessels hove into view, and in introducing valuable technological innovations they gained considerable influence over local chiefs as advisors. Historians have not always looked kindly upon them. Described as the "founders of dynasties and the bane of missions, governments, and more conventional colonists" [11], perhaps the most remarkable is William Mariner, a cabin boy who was captured by a Tongan chief in 1806, but went on to develop a militia army which played a pivotal role in enabling the conquest of all the Tonga islands. He became a powerful chief, a celebrated warrior and an influential landowner in Tongan society, before returning to London in 1811 to become a stockbroker! His book *An Account of the Natives of the Tonga Islands, in the South Pacific Ocean* is apparently still considered a vital text in understanding early Tongan culture. [12]

By the end of the nineteenth century, except on the most remote islands, the beachcomber was often regarded as little more than a hobo of the Pacific. There were exceptions, the most celebrated example being E.J. Banfield, who, with his wife and housekeeper, lived on Dunk Island (six square kilometres, and four kilometres off the Queensland coast) from 1897 until his death there in 1923. The first of his four books, *The Confessions of a Beachcomber*, published in 1908, gained him an international reputation reflected in the fact that he was reviewed in the *Times Literary Supplement* and his obituary appeared in the London *Times*. [13] In keeping with beachcomber traditions, Banfield was not a solitary, and maintained strong links with mankind. A steamer called regularly at the island, and his "Rural Homilies", newspaper articles and correspondence indicates he was very well informed about world affairs.

In contrast to Banfield – and distinct from the castaway

who is shipwrecked, cast into the sea and onto a shore – the voluntary maroon seeks out an island purposefully, taking advantage of its isolation to embark on a kind of social experiment sealed from mainland influence. Crusoe himself was never happier than when he anticipated his rescue, but his island life has inspired a myopic obsession in the pursuit of the notion that a future life on an island may liberate us from the travails of the present life on the mainland. Whilst the number of people who have put this ideal into practice is small, there appears to be no shortage of people keen to emulate Alexander Selkirk if the opportunity arose.

Examples are numerous and diverse. When the last few remaining families evacuated St Kilda in the Outer Hebrides in 1930, after years of a precariously marginal existence that demanded they adapt perfectly to very limited resources, there were some 400 applications to the estate owner from people wishing to carry on a tradition of human habitation.

In 1961, Bernard Stanbury placed an advertisement in the London *Times*, asking for volunteers to establish a new society on an island. Property would be held in common ownership, living would be communal with some privacy in family units, education would be non-competitive, children would be brought up by the best "upbringers", and fellowship and love for all was to be the foundation stone of the society. Within a week there were 150 applications, and the number eventually swelled to more than 5,000, of whom he was able to choose twenty-six. By July 5, 1964, the *Sydney Morning Herald* was reporting that Stanbury had been unsuccessful in buying Middle Percy Island in the Great Barrier Reef, that Harvey Island in the Pacific Cook Islands was another possibility, and that his group was continuing to seek funding. Likewise, a two-inch advertisement in *The Observer* newspaper for two "castaways" to act as personal assistants to anthropologists in the Tuomoto Archipelago in French Polynesia elicited over 300 responses.

Such was the media attention given to eighteen-year-old Martin Popplewell, who planned to cast away with Helen, his "Girl Friday", that in January 1989 the mass-circulation *Daily Mail* in Britain devoted half its front page to the story. Voluntarily marooned – not cast away – on Patangeras, which is within Ulithi Atoll in the Pacific's Caroline Islands, and inspired by the Hollywood movie *Blue Lagoon*, Popplewell's three and a half years of planning culminated on an island strewn with rubbish from World War II, inhabited by thousands of rats, and infested with mosquitoes. They argued bitterly, and Girl Friday left after a month to sell her story to the tabloid press, describing her Crusoe – twelve years younger than her – as a man obsessed by an adolescent dream.

Licking his wounds but still undeterred, he retreated to nearby Mog Mog, from which he was able to persuade his childhood sweetheart, Rachel, to share nearby uninhabited Dorenleng. Their relationship remained "amicable but volatile" until salmonella and amoebic dysentery forced them to abandon their "test of survival". However, twelve years later Popplewell, now a television presenter and journalist, was able to return to Dorenleng with Rachel and use his original video diary as the basis for a feature film, *The Real Castaway*, which at least added an edge of reality to the increasingly popular mainstream castaway docudramas.

Castaway 2000, *Shipwrecked* and *Celebrity Love Island* have been prime-time television successes, combining the almost

instinctive desire of so many people from so many backgrounds to "make a new society on an island" – shared only
with a reality television audience! *Castaway 2000*, set on
Taransay in Scotland's Outer Hebrides, intended to "build
a new community for the new millennium", and what
started as a middle-class documentary finished as tabloid
fiction. When it was revised in 2007 on Little Barrier Island,
Australia, it was to be a twelve-week rather than one-year
social experiment, thus reflecting the accelerating trend of
instant gratification in post-*Big Brother* reality television
since 2000. When the idea was resurrected yet again in
2014, there were thirteen purportedly ordinary men left
on a Pacific island to fend for themselves, to "fight for their
existence", so it was said. It was to be for just a month,
and within a few days the television documentary *The
Island* was accused of fakery: four of the "ordinary men"
had allegedly worked in dangerous environments before;
the contestants were supposedly supplied water in a pre-
placed rubber-lined pool; and it was claimed that the two
crocodiles had been introduced for effect by the producers.
Bear Grylls, the programme's host, rejected the allegations.

In 2017, one could watch actor Robson Green fulfilling
his "dream of a lifetime" emulating Robinson Crusoe
on "untouched since the beginning of time" (sic) North
Guntao in the Philippines. He landed with a film crew,
stayed for a week in total (the time interrupted by suspected
gastroenteritis necessitating him being shipped off for treatment on a neighbouring island), and even the fragments of
serious intent were reduced to poor entertainment.

But from these examples alone it would be a mistake
to assume the allure of the desert island is fading. Since

Roy Plomley first introduced BBC wireless listeners to *Desert Island Discs* in January 1942 from a bombed-out studio in London's Maida Vale district, the programme has become a broadcasting institution, seemingly impervious to change in a business that has seen its shibboleths tumble in recent decades. As one critic describes it: "An invitation on to *Desert Island Discs* is a measure of one's ranking in the collective consciousness, up there with getting an OBE." [14]

Since its outset, guests have been invited to select eight pieces of music and, from the late 1950s, they have been able to select a luxury item, and a book or set of volumes. These were to accompany them when they were, theoretically, marooned. Roy Plomley never went to a desert island, not even to do research. The BBC once tried to send him to one but he refused, complaining of the dangers of poisonous fish, poisonous berries, and crabs. Quite reasonably he did not wish to risk his life even if it did boost listening figures. Whilst the guests are usually asked how they think they would "get on" alone on their island, the programme's affinity to island experience seems to be the notion that using a desert island as a stage enables interviewees to look back on their lives in a "nature in the raw" kind of way. And there is the assumption that the question of how (urban) people in the public eye would cope with the isolation of Robinson Crusoe has universal appeal. The formula seems to work even if many of the celebrities are honest enough to admit early on in the interview that their chances of survival would be minimal, leaving time for dialogue that is more often gently thought-provoking than particularly revealing.

Quite unlike most of the castaways on *Desert Island*

Discs, Gerald Kingsland was a "man who wanted to be Robinson Crusoe", whose advertisement in a London magazine sought "couples and single people ... to colonise uninhabited tropical island". [15] He may have claimed that he wanted to emulate the famous castaway, but in fact he had no desire to be shipwrecked and thrown onto a rocky shore. Rather, like Alexander Selkirk, he was marooned by request three times as he strove to make a new life on Cocos Island (about 550 kilometres west of the Pacific shore of Costa Rica), Robinson Crusoe Island (the largest of the Juan Fernandez Group, about 675 kilometres west of South America), and Tuin island (in the Torres Strait, 110 kilometres from Papua New Guinea). He succeeded to the extent that he survived, but by his own admission he met with failure either because the islands were too frequently visited for his new world to flourish – utopias seldom survive out of isolation – or, as on Tuin, he and his companion, Lucy Irvine, were ill-equipped to survive in a harsh environment where life could be reduced to not much more than relentless acts of endurance.

A confessed self-publicist, his writing is laboured and sometimes of dubious authenticity as he admits trying to create the character role of a book he was intending to write. Irvine's account of her co-habitation with Kingsland on Tuin is more insightful, but ultimately both of them were embarking upon "adventures" from which they hoped to capitalise financially by narrating tales of derring-do. And, more importantly, they fell prey to the way in which small islands act to expose, magnify and render untenable what may broadly be described as "difficult personality issues" – issues that in other less intense environments we can more or less get along with. There is consolation of a sort for Kingsland at the end for, having been repulsed by both Tuin and Irvine, he finds solace in being respected and valued by neighbouring islanders less for emulating Robinson Crusoe than for his skill as a mechanic.

If Tuin reduced its voluntary maroons to unanticipated "relentless acts of endurance", they were pre-empting a modern coterie of "extreme adventurers" who purposefully seek an island experience that will test the limits of their physical and emotional strength. Rockall, with a land area of just twenty-five by thirty metres and peaking at seventeen metres above sea level, lies 400 kilometres west of Scotland's Outer Hebrides. In summer it can be subjected to sea spray for days on end, and winter seas can overwhelm the island completely. In 1985, ex-SAS and Parachute Regiment Officer Tom McClean spent forty days there, living in a plywood box 150 by 120 by 90 centimetres. He insisted his aim was geopolitical, to be in occupation long enough to reinforce UK claims to Rockall, which is otherwise claimed by Ireland, Iceland and the Faroe Islands. In the annals of extreme adventure competition, McClean's solo record was not broken by a group from Greenpeace who occupied the island for forty-two days to draw attention to the destructive potential of oil exploration in the area. However, the record changed hands in July 2014 after Nick Hancock lived there for forty-five days in a "reinforced plastic survival pad" that looked suspiciously like a septic tank.

I find it difficult to celebrate these activities. McClean seemed to be indulging in the macho twilight of imperialistic sabre-rattling and Hancock was more than anything else determined to impress the world with his solar panels, wind turbine, satellite phone and reinforced laptop with which to blog, text and Skype his way into public consciousness.

A small island does not free us from the psychological baggage we ferry ashore; rather, island life provides only the briefest interlude before it starts to rummage about in one's persona, seeking weaknesses to exploit. Bad history will surely muddy the waters of the future.

The Story of Tom Neale

Something that is delightfully free of psychological confusion is *An Island to Oneself*, published in 1966, which is Tom Neale's account of living alone on Anchorage Island in Suvarov Atoll. [16] Our understanding of his experiences can also be complemented, in a limited way, by Robert Dean Frisbie's account of living on the same island during 1944, by James Rockefeller Jnr's account of meeting with Neale on Anchorage in 1957, by James Simmons who talked with Neale in Rarotonga shortly before Neale's death and who visited the atoll in 1980, and by anecdotes and opinions of people who live on Rarotonga. [17]

Anchorage is approximately one kilometre by 300 metres, and its maximum elevation is four metres. It is 320 kilometres from Nassau Island, 385 kilometres from Manihiki, and 930 kilometres from Rarotonga, the main centre of the Cook Group. And it is 500 kilometres from the nearest shipping route. The atoll consists of about twenty tiny islets scattered across a shallow lagoon and encircled by a coral reef, which is approximately fifteen kilometres by fifteen kilometres in size. Neale's island exhibited just the qualities that Fletcher Christian and the remnants of the crew from the *Bounty* sought on Pitcairn Island 150 years before: uninhabited, unvisited and without a suitable harbour for shipping. The earliest "discovery" of Suvarov Atoll by Europeans is unknown, but Russians visited it in 1814 and they appear to have considered it unimportant. Like many isolated islands remote from trade routes, Suvarov has a romantic and unlikely history that belies its size, featuring castaways, coastguards, buried treasure and pearl poachers. Rockefeller Jnr. claimed that $10,000 in Mexican money has been unearthed there.

Thereafter, its history proceeds more conventionally. Suvarov was annexed by Britain in 1888 as the possible site for a cable station, and like the rest of the Cook Group, it came under New Zealand administration a year later. It was cultivated as a copra plantation (where oil is obtained from dried coconut kernels) prior to World War II when Anchorage was occupied by two New Zealanders, and their three Cook Islander assistants, who fulfilled the role of "coast watchers" and radio operators. The island has a long history of being used by pearl divers from Manihiki. For Tom Neale, the leftovers of human habitation included an overgrown shack and storehouse, tanks for collecting water from the roof with a total capacity of about 5,000 litres, the remnants of a garden, some five feral pigs that would threaten anything he planted, a wild flock of hens that could be more useful, a broken boat and a damaged pier.

Neale spent fifteen years alone there during three periods of residence. From October 1952 to June 1954 his solitude was broken by just two visiting yachts, and from April 1960 to December 1963 six yachts made Anchorage a port of call. July 1967 to May 1977 was a "busier" period as the reluctantly famous "Hermit of Suvarov" attracted more of the world's attention.

Born in New Zealand in 1902, he apprenticed as a naval engineer and for four years travelled the Pacific until he bought himself out in 1924. He did odd jobs and worked on inter-island trade ships, and after a brief return to New Zealand in 1928, he lived in Moorea, Tahiti, until 1943. He was the kind of rootless figure who could have been a beachcomber a hundred years earlier, and like beach-combers he knew how to make and build things, he was an imaginative improviser, and he was self-contained.

In 1943, he took the job of a relief storekeeper in Rarotonga in the Cook Islands, where he met Robert Dean Frisbie – variously described as writer, traveller, vagabond, father of four children – who had spent much of his life on isolated atolls in the Pacific. He fired Neale's imagination with his stories, culminating in Frisbie telling him that "Suwarrow is the most beautiful place on earth, and no man has really lived until he has lived there". [18] Neale realised that Frisbie had presented an image that had lodged in the back of his mind for a long time – a beautiful, remote island where he could make a home. In 1945, when the "coastwatchers" were being reprovisioned, he visited Anchorage on a trading schooner, and the peace and solitude of the place confirmed his desire to live there. But it would take him seven years to achieve his goal, and by late 1952 – having failed to gain permission previously – Neale persuaded the new resident commissioner in the Cook Islands that he could occupy Anchorage.

Neale took with him what he considered to be the bare essentials for living on an atoll: kerosene, soap, matches, knife and machete, fish spear, blanket, shoes for walking on the beach and a few clothes; and, to make things easier, tea, coffee, sugar, flour, baking powder, lard, beef, rice, tobacco, butter or oleo, salt and condiments, file, hook and line for fishing, and reading matter. He also took a number of tools, material and seeds necessary for making a garden, and some additional domestic articles, including a pressure lamp. He considered that the major omissions were a wood stove, properly barbed spears, a sieve for making topsoil, and caulking material for the broken boat.

The physical environment – the island is twelve degrees south of the equator – is luxuriant and quite typical of moist atolls of the South Pacific. The vegetation cover consisted then of a thicket of young coconuts, pandanus palms, tauhuna, gardenia and hibiscus, as well as mature coconut, tamanu and mikimiki trees. Wild edible produce included pawpaws, breadfruit and bananas. Neale caught fish without difficulty, and at the right time of year there were terns' eggs for the taking. He maintained an adequate supply of water as there were short tropical storms even in the dry season. The existing cabin was in a well-shaded position, and he repaired it to a comfortable condition. He was able to make a garden in which land crabs, wild pigs and a shortage of insects for pollination were a greater problem than soil fertility.

This situation may be compared with that of Kingsland and Irvine, who wanted to live in isolation for a year on Tuin. This island off the coast of Queensland has no permanent supply of water, there is an extended dry season, and the sea harbours poisonous coral and stone fish that make fishing difficult and dangerous. Kingsland and Irvine equipped themselves with two kilograms of dried beans, one of porridge and four of rice, two packets of dried fruit, and just a few other provisions, which they wrongly anticipated would sustain them until their garden became productive. They also had an axe, a machete, a spade and fishing tackle. They lived in small tents more often than in a derelict corrugated iron shack.

Conversely, Neale, who was fifty when he started living on Anchorage, was experienced living in the tropics, was adequately equipped according to a well thought-out plan for living in a limited but reasonably fertile environment, and he liked his own company to the extent that he could be confident that solitude would not overwhelm him. His pre-eminent concern was whether he could withstand a hurricane on an island with a maximum elevation of barely four metres – something that his predecessor Robert Dean Frisbie had experienced with his young family with near-fatal consequences.

Neale's writing is matter-of-fact, like the man himself, but it is never stilted, and it avoids the primitivism and romanticism that pervades much Pacific island writing. Most remarkably for somebody without previous writing experience – or perhaps because of this – he exhibits a wonderfully innocent sensitivity and unselfconsciousness. He may have been helped along the way by the British journalist, travel writer and novelist Noel Barber, who visited Neale during his second stay on Anchorage, and who encouraged and assisted Neale to write. Barber wrote the introduction, but the extent to which *An Island to Oneself* is "ghostwritten" we will probably never know; and anyway, it is the descriptions of island experience that are important, and they have a profoundly authentic ring to them. They describe a man and his island, the writing of which could not be separated by a third party.

Unlike Frisbie, who planned for just a three-month stay, Neale did not anticipate a departure date. It is implied in his book that he would stay as long as old age did not incapacitate him. As the sailing boat that brought him to Anchorage departed he admitted that "over the years I had imagined this moment dozens of times, often wondering what sort of emotions I would experience at the actual moment of severing my last contact with the outside world. I had imagined I might be a little despondent and had thought too, there might be a sudden surge of almost frightening loneliness. But now the schooner was leaving, I felt nothing but impatience that the ship took so long to get under way. ... Once she was far enough away, I took off my shorts and waved them in a symbolic farewell. From that moment onward I never put on those shorts." And so Tom Neale, dressed in a loincloth, symbol of freedom-at-last, set out immediately to explore *his* island.

The dimensions of his world are defined as "roughly tongue-shaped, and [it] measures only three hundred yards at its widest point. I could take most of it in at a glance as I stood on the beach." But in this initial exploration he is keen to impress that this is no circuiting the island for a pleasure stroll; no, he is afoot to "find out where the best coconuts were growing, discover the whereabouts of the best topsoil for my garden, [and] examine the shallows with an eye to the best pools for fishing."

And this workmanlike, pragmatic approach to trans-forming the island into a productive unit characterised his first occupancy of Anchorage. He was capable, resourceful and energetic, and determined to develop the island according to a well-conceived plan. Like Defoe's Crusoe and Tournier's Robinson during the initial period of occupation, preparing for an uncertain future sometimes took the form of a work obsession. He constructed an oven and a network of paths from volcanic rock. He captured and penned the wild chickens and killed the feral pigs. He fished, trapped lobsters and collected seabird eggs. He learnt how to pollinate his plants by hand since there were no bees to do the job. And so he was too busy to feel lonely.

It was not that he was unaware that he was bringing old habits with him to Anchorage: "My dislike of cooking (only because it wasted time) amounted almost to a phobia at first, because I could not really adjust myself to the tempo of this new life, the fact that I did not need to hurry. Instinctively, I wanted to get any job done as quickly as possible, and at times I would be spurred on by melancholy thoughts that I would never get my garden started or build a run and raise a fowl population."

There were good reasons during the first few months to lay down quickly the infrastructure of his island existence. The approaching rainy season could render these activities impossible, and an unsheltered home would provide exposed and demoralising conditions. And, most importantly, the hurricane season was due to begin soon after his arrival. He knew from Frisbie that a hurricane could combine with a storm surge to sweep waves right across Anchorage; and the evidence of the 1942 hurricane – a tangle of uprooted trees, gaping holes in the reef and a number of islets washed away – was grim evidence of destructive power even ten years after the event.

He describes that, in the process of being single-mindedly busy and in his endeavours to establish himself on the island, time just seemed to disappear: "I slipped into the routine that was to become my life [waking in the morning] thinking how lucky I was to look forward to a day which was going to bring me nothing but satisfaction." And the days would end in ceremonial fashion when "if the weather were fine, I would brew myself a bowl of tea and carry it down to the beach. There I would sit with the faint sigh of the trade winds rustling in the palms which bent in the canopy over my head. Sometimes I would light a small fire to cook the cat's supper. ... On some evenings the air would be so still I could hear my own breath."

However, an earthly paradise this was not. He contracted fever; and he suffered what he describes as "a highly emotional phase ... caused by the total lack of tobacco, complicated by my hunger for meat and a rather worrying knowledge that I could no longer stand the taste of fish". This set him walking up and down the beach "in moody, sullen anger". Yet he was reluctant to kill anything unless it was really necessary. In a remarkably sensitive passage he describes taming a wild duck and the "murderous thoughts" that tormented and obsessed his dreams. His solution was to refuse to hand-feed the bird lest temptation got too much for him, but the duck then refused to eat the food he left on the ground. Eventually the bird disappeared, leaving Neale feeling "infinitely stale and dreary". He was obliged to kill the feral pigs not because of his craving for meat, but because they were ruining his every attempt to establish a garden. But he describes the aftermath as producing an overwhelming sense of melancholy as all at once "I was no longer a hunter, I was just an old man of fifty-one, alone on an atoll. I walked home slowly, deciding that I would bury the animal the following day. Nothing on earth could have induced me to eat any part of it."

Tom Neale, the sensitive environmentalist, describes his hobby of coconut planting on the small reef islands as repaying the debt of happiness he owes Suvarov. Coconuts would normally only grow on the rare occasions when nuts, blown by storms, drifted across the lagoon and reseeded themselves. And it was while he engaged in reciprocating what the island had given him that he suffered a paralysis in his back. With incredible good fortune, his life was saved by the arrival of only the second sailing boat in twenty-one months.

Neale's response to his injury, his departure to Rarotonga, and his eventual return to Anchorage six years later in 1960, reflect the profound bond of attachment he had made with his home island. He describes lying on the beach of an adjacent reef island and making the enormous effort required to reach safety as "not so much an instinctive sense of self-preservation, as a desperate craving to reach my shack. It was my only home and I *had* to reach it. ... [It was] this overwhelming instinct to be on my own island." And like Tournier's Robinson, Neale's deep-seated love for his home was reinforced by the realisation of imminent separation from it. This compares with the detachment evident in his rescuers, for whom "the [island] experience would never have any deeper significance than a wonderful interlude; a nostalgic memory. ... To me, in sharp contrast, the island was *not* an adventure, it was something infinitely bigger – a whole way of life; and so, if I had to leave Suvarov, I knew it was vital I should spend my last few weeks alone on the island."

Meanwhile his "staunch friends" accepted his wishes and sailed off to deliver his request for a ship to evacuate him; but when the Manihiki schooner arrived he admitted to a desperate desire to turn back the clock, to find excuses to stay on his island: "I just didn't want to leave. I knew, with a dull feeling of despair, that the last thing I ever wanted to do in life was to leave."

Six years later, and despite New Zealand government disapproval which made it very difficult to charter a boat that would divert to Suvarov Atoll, he was back: "I returned because I couldn't keep away from the place – my reasons for loving Suvarov have always been as uncomplicated as that." Indeed, the siren call of Anchorage seems to have been so strong that, although he does not mention it in his book, he felt no compunction about leaving his wife and children behind in Rarotonga.

But he appeared eager to make some changes to how he lived his island life. So, even as he was revelling in redis-covering meaningful routines for daily life, he recognised that his activity need not be determined by the embedded attitudes and behaviour that characterised the early part of his first stay. Like Tournier's Robinson after the destructive explosion that ruined all the gains of his hard labour, Neale realised that in long hours of hard work "[I had] failed when I reached Suvarov the first time to put into practice the lessons I learned during half a lifetime in the South Pacific. ... I had been so proud of my island that I wanted to do everything in a rush. And so, in a curiously ironic way I had unwittingly imposed on the timeless quality of the island the speed and bustle of modern cities which I had been anxious to escape."

By his toils he had behaved like the archetypal Western colonist: he had developed his island into a neat productive unit. Even on a tiny spit of sand he had managed to "complicate" a life that was intended to be "simple". He recalled that Defoe's Crusoe had built huts several kilometres away from his stockade, and so he decided to build a comfortable cabin on the nearby island of Motu Tuo to which he could retreat and so avoid his proclivity for hard labour.

Neale was healthy and energetic, and the notion of growing old and dying alone was likely something he did not wish to dwell on. But he could hardly have foreseen the events that brought his second sojourn to an abrupt end in December 1963 when a group of pearl divers arrived from Manihiki intent on staying for a couple of months, perhaps every year in the future. Neale observed his fresh-water supply shrinking, litter strewn about "his" beaches, and realised his much-valued sense of tranquillity was a thing of the past. How fragile had been his faith in a permanent peace. How suddenly Heaven became Hell. And so he was on his way to Rarotonga at the first opportunity. [19]

Thereafter it seems Neale had no intention of returning to Anchorage. He was always able to pick up his job as a storekeeper, and seemed equally at ease in resuming life as a family man. He intended to write his island memoir, and he may have even had it in mind to "settle down". An Island to Oneself was an immediate bestseller in New

Zealand and Australia, and soon achieved steady sales in other English-speaking countries. Reliving his experiences may have made him feel nostalgic. Three years may have been longer than he could bear to live in one place – unless it was on Suvarov Atoll. Royalties from his book made it less imperative to do a job that Neale must have found too easy on the head and too hard on the heart.

People read his book and were enchanted. The oceanographer Henry Stommel wrote that, "as an islomane, I find [the book] engrossing. It is like an enchanting dream: the lonely white beach of remote Suvarov Atoll, the sound of the distant breakers on the reef, the peace. And when I read in the census of the Cook Islands that in 1971 Suvarov still had a population of one, I wonder who he can be." [20]

What Stommel appears unaware of is that after the publication of *An Island to Oneself* Neale returned to his island for a third time, and that it is he who is the "population of one" in the 1971 census.

So again leaving his family, he had returned to Anchorage in July 1967, remaining for ten years. This time he worked with the pearl fishermen, and stayed on after they left. James Simmons, who visited Neale on Rarotonga shortly before his death, and who visited Suvarov in 1980, provides an excellent account of Neale's life during these years, and I am indebted to this in concluding Neale's story.

When Neale arrived on Anchorage for the first time in 1952 he brought with him basic essentials and not much else. He used just about everything and there were only a few items he regretted omitting. Fifteen years later, he landed with two boats, several drums of fuel, substantial building materials and about forty boxes of personal belongings.

And it was not only his perceived needs that were different. Neale had shared with his readers his love for the beauty of solitude, and now they wanted to share his island, to feel what it was like to be in paradise – but to leave easily when the pleasure waned. This voluntary maroon, who was most often described as a "castaway" or a "hermit", had become something of a celebrity in this corner of the Pacific, and a visit with him was a mark of distinction and prestige amongst fellow sailors. [21]

Neale may have regretted accepting the role of Postmaster, bestowed upon him by the authorities on Rarotonga who were looking to promote tourism in the Cook Islands – "The Last Haven on Earth" – and seeing Suvarov as a potential "stamp island". But the postmaster seemed almost overwhelmed by his duties and by his fanmail with its endless questions and requests for memorabilia. The lagoon began to feel crowded, Neale found it difficult to fall back into his old routines, and the yachts left behind newspapers and magazines. His life on the island had always been characterised by a narrow range of fundamentally significant immediate issues, but he now "found himself growing increasingly concerned about the month-to-month problems of the world. His letters through the 1970s are filled with expressions of concern. ... Neale had become a celebrity and not even the incredible isolation of Suwarrow could give him back his privacy or his peace of mind. His life had changed forever." [22]

Comfortably set up on the royalties from his book, twice visited by his daughter, but generally grumbling about his visitors, there was sadness and resignation in the realisation that his isolation and idyll were broken. Yet one wonders whether, despite his protestations to the contrary, Neale in his increasing old age may not have got some satisfaction in playing the wise old island man. Certainly he carried out a great service in enriching the experience of visitors, and their perception of the encounter could be entirely different from that expressed in Neale's grumblings.

Visiting yachtspeople described him following a busy daily "schedule" that was almost ritualistic, a habit that trailed off in the evening when he became a willing raconteur who appreciated an audience. [23] Kenneth Vogel, who lived in Samoa from 1974 to 1977, wrote to me that, "during the sailing season, Neale's little 'harbour' became a port of call of a number of sailing yachts. People would bring him presents, and he would reciprocate with fresh veggies and stories of the old days. It was very pleasant indeed; I remember some super evenings just sitting on the beach with him watching the sun set and being regaled with his musings."

In May 1977, a passing yacht found him seriously ill. The inter-island schooner was diverted to take him to the hospital in Rarotonga, where he was diagnosed with terminal cancer. He died at the end of November 1977. Suvarov is now a wildlife sanctuary. Such is its popularity amongst Pacific sailboaters that concern has been raised about the viability of bird colonies and fish stocks around the lagoon. And in 1985 officials started camping on the island until the beginning of the hurricane season in an attempt to stop visiting yachts from staying more than three days.

It is tempting to apply a simple psychological interpretation to Tom Neale's motives and experience: here was a recluse, a man who felt ill-at-ease in society, and who sought extreme isolation as a means of escape – an almost pathological reaction to the world at large. Certainly it is true that, whilst many may envy Neale the tranquillity of his life on Anchorage, very few would relish isolation with the casual appreciation that Neale exhibited. But there was nothing in his past, in his book or in what people have said of him to indicate that here was a recluse. Indeed, he admitted he had never been one for indulging in unnecessary discomfort and he does not fit the common conception of an isolate. The apparent ease with which he left his family behind (twice) to return to his island, and his explanation that they would interfere with his freedom too much on Suvarov, suggests that he was a man of his time rather than an unconscionable hermit. For this he paid a price – he was divorced in 1972.

We can only explore his motives now in a superficial way, but it would appear that there is little in his background to indicate that here was a man seeking an isolated island in order to escape from a lifetime of mainland anxiety. Unsurprisingly, given the success of his book, even forty years after his death, he is remembered locally, and not always entirely sympathetically. Indeed, according to one website, "many Rarotongan residents have anecdotes and opinions of him and it seems that his book, which ghost written, makes him out to be a much more reasonable fellow than he actually was. One person's opinion was that he was so cantankerous that an uninhabited island was the only place for him!" But there is no suggestion that he had the reclusive temperament of an isolate. [24]

It is important to be aware of Neale's positive associations with mankind in general in order to see that his island life, though unusual, was the product of a healthy personality doing something so many islomanes might aspire to, but only romantically in the landscape of their imaginations. His vision of living alone on Anchorage might seem romantic, but he went about achieving it in a wholly pragmatic way. This enabled him to fashion a domestic world through practical enterprise directed towards self-sufficiency. Suvarov Atoll was well chosen, conforming to an image of a Pacific earthly paradise that Neale was well aware of, and Neale was well suited to be there. Almost everybody who encountered him speaks less of his exceptionalism than of his ordinariness. Noel Barber, who was a crew member of one of the boats that called at Anchorage, described Neale as "modest, quiet, intelligent, with a wonderful sense of humour and a lively curiosity about events in the world he had left behind". And another visitor described him as "so … normal it hurts. He's just an ordinary, stubborn little shopkeeper and here he is living for eighteen months all by his lonesome." [25] And it is for the "unique normalness" that enabled him to live his dream that Tom Neale himself would perhaps like to be remembered.

Neale's life on Anchorage in a curious way supports the model developed by biologists to explain life on small, isolated islands. His island was very small – just a few hectares – and it was extremely limited in resources. Despite being vulnerable to hurricane seas, it had a climate conducive to outdoor living, adequate rainfall, a useful vegetation cover and the vestiges of earlier human occupation. Neale seems to have known himself well – in modern parlance, he was "comfortable in his own skin". He planned well, was practised in a number of skills which supported self-sufficiency and engaged with the challenges of existence in a logical and resolute way. He adapted well to his environment and was self-contained, but despite all this he was vulnerable from the day he set foot on his island. His well-being was predicated on the remoteness and isolation of his island home, and as soon as this was broken by external forces – by the pearl divers arriving and setting up camp and then by his successful book attracting visitors – so his life on the island could never be the same.

And as we shall see in the next chapter, this fragility and vulnerability to agents of change is an enduring theme in island history.

Chapter Three

A Fragile Geography

The Island

The fact that small islands can "rise out of nowhere and disappear overnight", the fact that this gives a tiny demonstration of how the Earth was made and how it will be unmade, and the fact that island production and destruction has the drama of immediacy and a disregard for the conventions of unimaginable geomorphological time scales – all this stimulates our imagination and curiosity. You blink and they are there, you blink and they are gone; and as a cluster of tiny islands disappears over the horizon, it is easy to believe that perhaps they were never there at all. So W. Somerset Maugham, in his short story "The Four Dutchmen", describes how little islands, "one so like another, allured my fancy just because I knew that I should never see them again. It made them strangely unreal, and as we sailed away and they vanished into the sea and sky it was only by an effort of the imagination that I could persuade myself that they did not with my last glimpse of them cease to exist." [1]

In Somerset Maugham's imagination, islands come into existence as they rise over the horizon and cease to exist when they drop behind it. And we apprehend this subjective response because it is uncomfortably close to the precarious objective reality that defines small islands. Their very existence depends upon a precarious equilibrium between creative and destructive forces so that their geography is one of impermanence and uncertainty. This state of contingency is manifest in temporaneous words like shoal, flat, sandbank, bar, spit and reef, all of which are "nearly" islands, or islands that are sometimes there and sometimes awash or submerged; while rocks, cays and keys have just about made it to form islets. Some cluster together for protection in groups, chains, atolls and archipelagos, but even this makes for uncertain security. And, whilst remoteness may protect islands from external influence, when change occurs – concentrated as it is on a small landmass – it can be more radical, more disruptive and more complete than on the mainland.

Island ephemerality has its origin in Plato's fourth-century account of the inundation of Atlantis, an event that has inspired a plethora of scientific and pseudo-scientific research and associated literature. The interest in the legend, or myth, shows no sign of abating. Plato located Atlantis in the Atlantic west of the Pillars of Hercules, but despite the area being scoured by numerous ever-optimistic expeditions, and despite some 2,000 books having been written on the subject, nothing substantial has been found. In the mind's eye of Atlantis researchers, it has been located in places as far apart and improbable as the Antarctic, the South China Sea, the Azores and elsewhere in the mid-Atlantic, the Caribbean, Cornwall, Indonesia, somewhere near the Gulf of Corinth and on Santorini, between Sardinia and Sicily, Troy, Bolivia, between Sri Lanka and India, Ireland, Crete, Malta, Tantalis in Turkey, Andalucia, the Black Sea, southern Finland, off the coast of Cuba, near the Bahamas ... and no doubt more.

An Island Born

Island-forming is not a process consigned to the furthest reaches of geological time. Rather, it exists in a contemporary time frame that is almost disconcerting in its vitality. There are about 100 volcanic islands that have formed in recent (post-Pleistocene) times. These have been in areas of tectonic instability, and most of the islands have been short-lived. Thera (now Santorini) had at least eight island-forming eruptions between 1600 BC and 1866 AD. Recent research suggests that the eruptions in 1600 and 1100 BC may have been responsible for a climate change that contributed to the decline of Minoan and Chinese civilisations.

In an area renowned for volcanic islands emerging out of the sea only long enough to be washed away, Julia Island appeared off southern Sicily in 1831. Concern that continuing volcanic activity could form a chain of new islands and destabilise Euro-African geopolitics, the eruption sparked a four-way dispute over sovereignty before Julia withdrew diplomatically beneath the ocean surface the following year. Survey work after eruptions in the same area in 2006 indicate that the potential island remains a discreet six metres below sea level. [2]

Elsewhere, during a devastating earthquake in Pakistan in 2013, it was reported that crowds of bewildered residents gathered near Gwadar to witness an island being thrust out of the sea, a sea that immediately began to wash it away.

But some tectonic islands have held on to life more tenaciously. Kavachi in the Solomon Islands has built itself above the sea nine times since 1950, and volcanic activity in 2000 was considered interesting enough to be included in British news bulletins. Likewise, the formation of Surtsey, off Iceland, attracted great media attention as scientists scrambled recklessly to make an early landing. In 2009, there was widespread island-forming activity just sixty-five kilometres north-west of Tonga's capital, Nuku'alofa, centred on a cluster of about thirty-five undersea volcanoes. Within this section of tectonic plate instability (part of the Pacific "ring of fire") an archipelago of 170 islands, with hundreds of square kilometres above sea level, was thrown up in just one week – long enough for fast-thinking entrepreneurs to set up a volcano sightseeing operation.

Likewise, on the other side of the world, global warming tourism is underway with the discovery in 2005 of a finger-shaped island in East Greenland's Liverpool Land, which had emerged from the ice sheet. Already an expedition company has promoted a "voyage of discovery" to be among the first to see this island, which has been named Warming Island. One imagines there could be many more to follow.

The Gironde river in western France has a history of shifting sands formed by a sediment-laden ocean current being slowed by water emerging from the estuary; but in 2000 a significant sand bar started to form, at first only visible at low water. By early March 2009, vegetation was observed for the first time, thus confirming it was not covered by water at high tide. An island was born.

Local people called it L'Île Mysterieuse, but it is also known more prosaically as L'Île Sans Nom. About double the size of a football pitch at high tide, it is more than ten times bigger at low. Its existence is precarious – early in 2010 it was cut in half by storms and the metre-high vegetation was stripped off. But it repairs itself. Ecologists, scientists, environmentalists and politicians are excited by this new piece of France – which means that its emergence has not been free of dispute. Scientists see an opportunity to study in a natural laboratory precisely how a new piece of land is colonised. Local politicians, perhaps sensing potential for ecotourism, tend to favour this. They recognise the value of an administration able to protect the island with regulations and landscape designations.

But already the appetite of local entrepreneurs has been whetted. There is a water taxi service from Royan, just six kilometres away, bringing day visitors from Paris and beyond, and sometimes there are more than 200 people on the island. The vegetation, what is left of it, is being trampled. There have been all-night raves, and the "beach party people" even have a "liberation front", which claims that people have been using the sandbanks for more than fifteen years.

The island is small and accessible, and it has emerged from the sea as a *tabula rasa,* fragile and vulnerable. As a new-found-land it has captured the public imagination, partly because of its scientific interest but also due to it being a natural stage where passions are compressed and intensified. It is possible that L'Île Mysterieuse will solve its own problems by disappearing, but it seems just as likely that, like L'Île Nouvelle, which appeared in the Gironde estuary in the eighteenth century and still exists, conflict will (g)rumble on. [3]

Not far away, in northern France, is Mont Saint Michel, a medieval abbey, and second only to Paris as a tourist attraction. The island (and I use the word guardedly, for it is, except in extraordinary sea conditions, connected by a causeway, soon to be replaced by a bridge) is no stranger to controversy. Victor Hugo, in 1884, exclaimed that "the Mont Saint Michel must remain an island. We must save it from mutilation." Since that time, when a dyke was built to the island and a local tidal river was redirected, mud flats have encroached to the extent that the Mont is only cut off at very high tides, most recently in July 2013 when it was enisled for the first time in 134 years. Experts believe it could be permanently connected in forty years.

What is at stake now is a €200 million, six-year scheme to reverse years of land reclamation by building a dam that will allow river and sea water to flush away three million cubic metres of mud and sand. This project, which it is anticipated will enable the Mont Saint Michel to be encircled by high tide up to ninety times a year, seems on the surface unprovocative; yet it has been the cause of political battles, court cases, strikes and threats by UNESCO to suspend the Mont's World Heritage status. As John Lichfield suggests in his excellent articles in *The Independent*, God and Mammon will have to fight it out for the soul of the abbey! [4]

The proposed development involves banning the parking of cars on the salt flats surrounding the abbey and developing infrastructure, particularly a car park and a road train on "the continent" – as the locals call the adjacent shore – which would bypass the lucrative tourist facilities of the local business community. The mayor of the island has claimed that the disputed parking fee will make the Mont available only to a wealthy elite.

Vulnerable by Nature

Island-forming can proceed at a pace that seems to defy the conventions of geomorphological time scales. But their destruction can be equally dramatic. Islands in the Caribbean and the South Pacific are especially prone to natural disasters such as hurricanes, cyclones and volcanic eruptions. In the Cook Islands, Toka was swept away by a tsunami in

1914, an event given added significance by the terrifying accounts of Tom Neale and Robert Dean Frisbie, both of whom survived hurricanes on Anchorage Island in the same group. Tuvalu was an island group of seven atolls until a series of cyclones in the 1990s completely overwhelmed one of them.

Gunnbjörn's Skerries, between Iceland and Greenland, which supported eighteen farms in 1391, were represented on a Dutch map of 1507 by the cryptic comment *totaliter combusta*. We may never know what happened to the farmers of Gunnbjörn's Skerries, but in recent times several islands have been evacuated due to volcanic eruption, including Niuafo'ou (Tonga) in 1946 and Lopevi (Vanuatu) in 1960. Between 1980 and 2000, several islands in the Mariana Islands Archipelago were evacuated due to violent volcanic activity. These include Anatahan, Alamagan, Agrihan and Pagan, which had a civilian population of about 650 when it was garrisoned by Japan in 1942.

A similar fate befell Tristan da Cunha, which lies in the South Atlantic, about 2,000 kilometres from St Helena and 2,800 from Cape Town. It claims to be the remotest inhabited archipelago in the world and has a population of about 270. There is no airport and no natural harbour. In 1961, a long-dormant volcano awoke to generate heavy earth tremors, causing a dangerous rockslide above the only settlement of Edinburgh of the Seven Seas. A huge fissure opened up behind the lighthouse, and the islanders were evacuated to another part of the island, then to Nightingale Island, thirty-five kilometres away. From there they moved to Cape Town, then to Britain. In 1962, after much agitation by the islanders to return, the Royal Society mounted an expedition to investigate the eruption and its effects on the island settlement and infrastructure. Some of the islanders returned that year, and almost all of the others in 1963. [5]

Small islands occupy limited space, with centres of population and economic production often in vulnerable coastal locations. There are intimate linkages with ecosystems, thus making impact-control difficult, which increases the chance of entire islands being made unproductive or even uninhabitable. For example, a stable coastline will protect the interior water table, but when there is coastal erosion and sea level rise, water supplies are easily salinated. There is a high ratio of coastline to land area, leaving islands highly vulnerable in general to a range of marine and climate influences with effects that will only be exacerbated by global climate change. According to the United Nations, of twenty-five small island states, thirteen are considered amongst the most disaster-prone countries.

The smallest islands do not generate their own climate patterns that might enable them to deflect hurricanes and cyclones. Some island ecosystems evolve to have some resilience to such events – for example, in particular forests that regenerate after frequent hurricanes – but this long-term ecological process is thwarted where there is human intervention in the form of agriculture, which is usually the case. Even then, some crops, like sugar, have a relatively high resistance to hurricanes, whereas a crop like bananas does not.

Hurricanes constitute the most significant natural calamity among Caribbean islands. In 1955, Grenada's biggest industry, nutmeg cultivation, collapsed after a hurricane. The banana plantations in St Lucia were destroyed in 1980, as were five million forest trees in Dominica. The direct costs of Hurricane Gilbert in Jamaica in 1988 were estimated to be $956 million, and that probably did not include contingent damage to the natural resource base. Media reports suggest that in 1995, Hurricane Luis destroyed seventy-five per cent of buildings in St Kitts and Nevis. [6]

It had been more than sixty years since a hurricane passed close to Montserrat in the south-east Caribbean, so long in fact that it was no longer considered a "hurricane island". Then, in September 1989, Hurricane Hugo swept across the island. It is estimated that a third of the buildings were destroyed, a third very severely damaged, and a third were less severely affected by the winds but inundated by the succeeding deluge of rain. The jetty alone, upon which so much depended for supplies, would take a year to rebuild. Economically significant "villa tourism" was wiped out in a matter of hours, something that the imminent arrival of "hurricane tourism" would, in the short term, do something to replace. World-renowned music studios were closed. The agricultural export trade vanished overnight. And the controversial "brass plate" international banking operation – some of which was being investigated by the British Fraud Squad over allegations concerning the laundering of drug money – was now nothing more than brass plate. The island's Chief Minister reckoned it would take fifteen years for Montserrat to recover completely.

But the island did not get fifteen years. On July 18, 1995, less than six years after the hurricane struck, volcanic activity started in the Soufriere Hills in the middle-south of the island, gradually rendering uninhabitable all but a third of the hundred-square-kilometre land mass. By August 1997, much of the infrastructure, including the Georgian-era capital Plymouth, the hospital, the airfield and the parliament building had been destroyed. Some two-thirds of the population has been forced to leave, most going to Britain or neighbouring Antigua and the total population, which stood at about 13,000 in 1994, has reduced now to about 5,000. Volcanic activity has been relatively quiet since 2010 when the remnants of the airport were buried in a lava flow, and some islanders have returned. There are plans to cash-in on the disaster – to harness geothermal energy, to mine ash and sand and to sell Plymouth as the "Pompeii of the Caribbean" to tourists. But with more than half the island an exclusion zone, with the economy stagnant during rebuilding, and with the spectre of a volcano still hanging over Montserrat, the fragile future of the island is the one thing that is not in doubt. [7]

Changing Climate and Ecology

In April 2007, *Time* magazine reported that the population of Carteret Island, to the east of Papua New Guinea, had been moved to higher ground on a new island. In the same month, the *New York Times* reported that some of the delta islands of the Indian and Bangladeshi Sundarbans region had been overwhelmed by Himalayan floodwaters, thus joining several other islands that have been completely submerged in the recent past. Increased run-off into the river systems as well as more numerous tropical storms and higher tides are almost certainly the cause. [8]

Throughout the world, not just the prospect but the reality of global warming is profoundly alarming for many low-lying islands and island states; and in places as diverse as the Orkney Islands, the cays of Australia's Great Barrier Reef, the Frisian Islands and the Florida Keys, the future is uncertain. The UN Climate Summit in Paris in 2015 agreed to seek a limit on global mean temperature increase to less than two degrees Celsius above pre-industrial levels. Taking the average of thirty-two models used in a recent International Panel on Climate Change report, it would appear that with an optimistic scenario the two-degree threshold will be crossed in about 2055, whilst a scenario based on present carbon dioxide emissions indicates the threshold crossed in about 2045. There are significant global variations – for example, the Arctic may warm much quicker – but for small islands in equatorial regions it seems likely that increases will be comparable with mean global warming. These temperature increases are consistent with future sea-level rises of between fifteen and thirty centimetres by 2050, and some scientists believe there could be a sea-level rise of between seventy-five and one hundred centimetres by the end of the twenty-first century. [9]

The factors associated with rising sea levels are inter-related, compounding in effect and far from understood, but small low-lying islands would certainly face the prospect of associated increases in the frequency of rainfall and humidity, a significant increase in storm conditions, wave energy and the destructive power associated with high tides, and altered ocean current patterns. This would threaten the survival of several island nations in the Pacific, including Tuvalu, Tokelau and Kiribati, whilst Micronesia, too, is particularly vulnerable. Archipelagos like the Maldives and Cocos (Keeling) in the Indian Ocean would face complete destruction of their infrastructure. Amongst the Alliance of Small Island States, which is gathered together in common cause, many would lose valuable coastal land and main centres of population would flood. The list of potential casualties is long: in the Atlantic – Cape Verde, Guinea-Bissau, Sao Tome and Principe; in the Caribbean – Antigua and Barbuda, Bahamas, Belize, Cuba, Dominica, Grenada, Guyana, Jamaica, St Kitts and Nevis, St Lucia, St Vincent and the Grenadines, Suriname, Trinidad and Tobago; in the Indian Ocean – Comoros, Maldives, Mauritius, Seychelles; in the Mediterranean – Cyprus and Malta; in the Pacific – Cook Islands, Fiji, Kiribati, Marshall Islands, Nauru, Papua New Guinea and islands in the Torres Straits, Samoa, Solomon Islands, Tonga, Tuvalu and Vanuatu; and in the South China Sea – Singapore.

In 2014, nearly twenty million people were displaced by natural disasters and, in common with the previous fifteen years, more than ninety per cent were weather-related.

In 2012, it was reported that the government of Kiribati was in negotiations to purchase twenty-three square kilometres of Vanua Levu, Fiji's second-largest island, 2,000 kilometres away. The land would be used to settle 500 or so farmers to grow crops for Kiribati and to mine landfill for sea defences. With many of the island nation's thirty-two atolls only a few metres above sea level, the plan would not provide an insurance policy for Kiribati's entire population of 103,000, which may ultimately depend on resettlement assistance from Australia and New Zealand. [10]

"Migration with dignity" is the hope of the President of Kiribati, but he believes the deal struck by 195 nations in Paris at the Intergovernmental Panel on Climate Change conference in 2015 will not save island nations who had been calling for "1.5 to stay alive". The undertaking to limit global warming to two degrees centigrade and the promise of $100 billion for a fund to assist vulnerable nations suffering loss and damage – these offer a degree of hope, but they are long-term commitments impossible to enforce when economic growth issues are difficult to predict and promises are easily broken.

The Maldives consists of 1,192 islands clustered around six atolls. Some 250 islands are inhabited with a total population of 400,000. The capital, Malé, is one of the most densely peopled places in the world. It would be under water if the sea level rose by just fifty centimetres, so it is surrounded by a three-metre-high wall, which cost £30 million to build and took fourteen years to complete. The highest point on the islands is 2.5 metres in altitude, and eighty per cent of the land area is less than one metre above sea level. In 2004, forty per cent of the landmass was under water after a tsunami.

About seventy per cent of the Maldives foreign currency earnings come from tourism. A resort typically consists of a single luxury hotel on a previously uninhabited island. There are nearly a hundred resorts like this, and another fifty or so being built. Plans to lease about thirty more islands for development have been heavily criticised over environmental concerns; but the government can only be pragmatic in exploring the very few options available to help fend off International Monetary Fund concerns over its £90 million budgetary shortfall. It is ominous too that, despite the tourist board's motto promoting the "sunny side of life", many of the new developments feature large, floating pontoons, and it could be that unless the rest of Maldives's population receives the same protection, the future looks more bleak than sunny. [11]

Even twenty-five years ago, at the United Nations World Climate Change Conference in Geneva, drastic measures were discussed, like the complete evacuation of Tuvalu, Tokelau and Kiribati to New Zealand. There are already 4,000 Tuvalese in New Zealand. Many of the remaining 10,500 live in densely populated Funafuti, the nation's main atoll, significant parts of which are flooded under half a metre of water at the highest high tides in February and September. In 2006–7, Tuvalu completed its National Adaptation Plan but, ominously perhaps, its author left for New Zealand soon after. As in the Maldives, their ministers have jetted around the world with the daunting task of encouraging the rapidly emerging nations, like China, India and Brazil, to reduce their emissions. Tuvalu's Prime Minister Apisai Ielemia has even admitted that reports by the International Panel on Climate Control made him feel that his country was fighting against the impossible. Equally alarmingly, it has been reported that most senior figures admit, if you probe, that they are formulating their own personal exit strategy. [12]

This vulnerability of small islands to natural disasters means that they have become a focus for research studies, and indeed ever since Darwin's visit to the Galapagos Islands in 1835 contributed to his initial formulations concerning evolution, islands have been recognised as excellent natural laboratories. Their numerical significance, and the variations in shape, size and degree of isolation, provide the necessary replication of "natural experiments" by which evolutionary hypotheses can be tested. Indeed, even hard-nosed scientists seem to have found the small island an intrinsically appealing study object, an appeal for biologists verging on the romantic. [13]

And what is true for biologists is true for anthropologists and social scientists, too. Margaret Mead and Bronislaw Malinowski made huge contributions to the field of ethnography through their pioneering work in Polynesia. And one small Irish island can perhaps lay claim to being one of the most intensely researched areas in the world. From the Royal Irish Academy's Clare Island Survey (carried out between 1901 and 1911) to their New Clare Survey initiated in 1990, biology, geology, archaeology, history, culture and geography have provided an interdisciplinary platform from which the island has been studied exhaustively. [14]

Indeed, scientists have sometimes had the unique opportunity to study remote small islands as a *tabula rasa* to trace ecological change. Clipperton Island is a minor French overseas territory, the remotest uninhabited atoll in the world, occupying just six square kilometres and situated 1,080 kilometres from the coast of Mexico.

Little over a hundred years ago, before it was settled for a guano mine, it was considered to have a more or less "plant free landscape". Palm trees and pigs were introduced, the latter feeding heavily on the millions of land crabs that had hitherto controlled any plant life. A low, weedy vegetation now began to creep across much of the island.

The pigs also decimated the internationally significant Masked and Brown Booby population, a problem resolved when Los Angeles County Museum ornithologist Ken Stager took a unilateral decision to shoot all fifty-eight pigs.

This enabled the crab population to recover so that it set about harvesting the new vegetation and feeding on every sprouting coconut.

Clipperton was on the way to having an ecology akin to pre-human occupation when, between 1999 and 2000, two large fishing boats were wrecked on the island. Rats came ashore. They fed extensively on the crabs, and this enabled a rush of vegetation and new plant species – including a creeping vine that was predicted to cover the whole island in five to ten years – which, together with an increase in sprouting palms, provided even more habitat for the rats. They were also feeding on the eggs and chicks of ground nesting birds, and Masked Boobies were facing a changing habitat detrimental to the open ground necessary for their flight pattern.

A Canadian marine research expedition visiting the island in 2016 observed a continuing and substantial decline in crab numbers, the "creeping flora" gaining terrain and coconut seedlings flourishing. No mention was made of the health of bird populations. [15]

And whilst the ecology has swung precariously between the conflicting consequences of human contact, other aspects of the landscape have not escaped attention. One's first impression looking at photographs is the enormous quantity of marine debris and "trash". Occupied by the United States during World War II and by French naval missions between 1966 and 1969, there are derelict, decaying makeshift camps, rusting steel ship hulls and all manner of equally rusty military, fishing and communication para-phernalia, decaying plastic nets and plastic containers, and neatly stacked live ammunition more than seventy years old. Pretty, it is not.

And so the Canadian report concludes: "When imagining far and remote places, isolation is usually perceived as a positive. We harbour images of virgin spaces, trouble free areas, dense vegetation, pure water, intact ecosystems and an abundance of predators underwater. Sadly, Clipperton resembles anything but this. In its case, isolation means that it is left to fend for itself and that no one is there to prevent the pillage of its resources." [16] As we shall see, this is a reality that could be said to apply equally to the Galapagos Archipelago.

A Dangerous Outside World

Oceanic islands are small land masses rising abruptly from the deep water of ocean basins or from submarine ridges. They are often volcanic in origin, have not been linked by continuous land to continents, and there is a limited range of fauna and flora, which is made up chiefly of forms with efficient long-range dispersal systems. Land vertebrates are characteristically few. Thus Tristan da Cunha has only five species of land bird, Henderson Island has four, Gough Island two, and Easter Island none.

The smaller the island, the more permanent the separation from mainland, and the greater the degree of isolation, then the more limited and distinct is the biotic community which colonises the island. But within these restricted communities there is less competition than on the mainland, so some species survive on islands whereas they may be excluded in more competitive environments. Genetic regression can occur, which tends to cause the loss of adaptive creativity, but because on small islands species are removed from related breeding stocks, the appearance of new characteristics will, if favoured by natural selection, evolve into distinct endemic species.

In the Galapagos, its penguin, marine iguana and more than half of the fifty-two bird species are unique. There are different species of tortoise on many islands, and there are thirteen unique Darwin Finches. Altogether, whilst the total of species is small – as one would expect given the isolation of the archipelago – forty-two per cent of the plants are endemic, as are seventy-five per cent of the land birds, ninety-one per cent of the reptiles, and 100 per cent of the mammals. It would appear that very few land birds over a period of several million years survived the enormous journey to Galapagos, but those that did were able to exploit a new world with few of the predators, diseases and competitors which acted as constraints facing their ancestors on the mainland. This provided opportunities for genetic variations to occur, aided by natural selection as new species were formed as they adapted over time to the unique environment. And this totality of new plant and animal species, through their interaction, formed new ecosystems that are themselves unique to particular islands. [17]

But there is inherent vulnerability, something Jacques Cousteau described as a "fatal duality". With isolation, a native population can become mentally and physically defenceless. A community limited in diversity, comprised of species that respond poorly to competition and which are uniquely adapted to a very particular environment, may demonstrate good *internal* stability within the island biotic community. But its fragility is exposed when the condition of separateness is broken by the introduction of *external,* non-endemic species and by processes of human exploitation. [18]

The Island

Again, Galapagos is an excellent example of how the "fatal duality" can work out. Although the first record of the archipelago dates from 1535 when Tomás de Berlanga was relieved to find a little drinking water in a land that otherwise appeared to be showered in stones, the islands merited little attention until about 150 years later. Buccaneers used them during the second half of the seventeenth century, taking advantage of water and timber supplies, and William Dampier, in particular, recorded abundant food supplies, especially in the form of a giant tortoise, the succulent meat of which could feed many men. When whalers arrived a century later, they found in the tortoises a source of fresh meat that stayed alive in a ship's hold for months without water or food. Although they had no interest in settling the land, their activities ensured that the giant tortoise became extinct on Floreana and Santa Fe (although some tortoises appear to have been "rediscovered" in 2011) and the population on Española became severely depressed. By the 1860s, sperm whales were scarce, and the fur seal was considered close to extinction by the early twentieth century. The sole surviving giant tortoise from Pinta Island, "Lonesome George", died in 2012.

With increasing numbers of ships visiting the islands, those highly efficient predators rats and cats were released, as were goats that soon formed large feral herds. By the mid-1990s, there were an estimated 10,000 feral goats on Isabella alone, threatening the survival of the giant tortoises. After years of debate, systematic programmes of eradication were implemented, and by 2006 the main islands were free of goats, donkeys and pigs, according to the Galapagos Conservancy.

In 1832, just three years before Darwin arrived on HMS *Beagle*, Floreana was settled, Ecuadorian sovereignty was established over the islands, and there began a process of human exploitation that has continued ever since. During the next hundred years, when three other islands were settled, cattle, dogs, chickens and pigs were introduced, as well as many plants (both agricultural and ornamental), which formed their own competing ecosystems. The settled islands were cleared to make way for houses, cattle, subsistence agriculture and plantation crops like sugarcane. But economic development and population growth was slow and faltering during the first hundred years and more. Distance to markets was great, on poor ground the islanders did not cultivate anything that was in short-supply elsewhere, and there was little wealth to generate internal trade.

Like so many nations that have offshore islands, Ecuador chose to use Galapagos for a penal colony, and to export workers from the mainland who were often treated little better than slaves. An atmosphere of simmering discontent meant that the production of hides, coffee, sugar and salted and dried fish for export was often interrupted by resentful workers refusing to work. Even the Norwegians, who in 1926 colonised Santa Cruz in anticipation of setting up something of a utopia far away from their own depressed homeland, soon gave up, faced with rocky soil, uncertain rainfall and poor organisation.

In 1950, the population was less than 1,500, but in the late 1960s increasing wealth in many "developed countries" meant that so-called "green tourism" was attracting wider sections of society. A new industry was created, depending not on the productivity of the islands as measured in more traditional activities, but only limited by the interest that could be generated in the tourist market. In 1970, with the Galapagos population less than 4,000, the annual tourist trade accounted for about 4,000 visitors. As of 2017, the population has increased eight-fold to about 32,000, and tourist numbers have increased more than fifty-fold to 225,000. In 1970, cargo vessels docked three times a year. Now four cargo boats and more than thirty jets bring goods and passengers every week. It has led, inevitably, to the introduction of non-native plants and insects, as well as plant, animal and human diseases erstwhile unknown.

The tourist industry, as it almost always does, has created inequalities in wealth and expectations. The archipelago earns approximately $100 million for the Ecuador economy, but it is estimated that some eighty-five per cent of this is exported to the mainland. Dr Godfrey Merlen has lived on Galapagos for more than forty years, where as a project scientist he has made a very significant contribution to natural and social history. He has witnessed wealthy visitors mingling with a large influx of untrained people from the mainland, who have "no roots in the Islands, no loyalty to their lands and little, if any, understanding of the special significance of the Archipelago. ... Their dreams shattered, disillusioned folk roam the streets of Puerto Ayora, unable to find work, yet unable to find passage to return to the mainland. Drinking and drugs are common. As for interest in biology, the subtleties of evolution? That, they would say, is for the wealthy, who have nothing better to do. ... Any natural resources are regarded as a means to immediate financial reward. ... Today Puerto Ayora is a typical tourist town. Its sister villages on San Cristobal and Isabel are jealous of its economic success. ... Behind this facade lie block after block of low-quality buildings. ... Here live the fisherman, tough, independent, suspicious of science, mistrustful of authority, accused of destroying the marine resources of Galapagos, and well aware that they are at the bottom of the economic heap." [19]

It is perhaps unsurprising then that marine resources have been under relentless pressure. For example, the growing trading influence of South-East Asia promoted a demand for sea cucumbers and shark fins, and so a highly lucrative fishery sprang up. There was neither regulation nor monitoring, no government department taking responsibility, and there was no definition of the protected waters around the islands and what activities were permitted therein. Any attempt at setting quotas was met by irate fishermen claiming poor people were being denied access to valuable resources.

The conflict threatened the conservation of biotic communities, just as it exacerbated the divisions that ran through the Galapagos human community. In 1997, a Special Law was presented to Congress, which would define appropriate human population levels, define protected waters in the archipelago, define mechanisms for generating finance for conservation, place a ceiling on tourism development and create a ministry of environment and an administrative body to seek conflict resolution in the community. It was first blocked by industrial fishing interests, but passed a year later.

The Special Law recognised that introduced species and the misuse of marine resources were the principle obstacles to the coexistence of people and the island's natural fauna and flora. This recognition has led to the Galapagos Marine Park being declared in 2001, including ninety-seven per cent of the archipelago and providing a sixty-five-kilometre offshore reserve boundary where only tourism and artisanal fishing is permitted.

But such measures have not been enough to stave off the destructive impact of Cousteau's "fatal duality" – the vulnerability to outside influences of island communities formerly protected by their isolation. In particular, the human population level is extremely high, and it remains to be seen whether a registration scheme will work to regulate numbers. Commercial fisheries, based 1,000 kilometres away on mainland Ecuador, continue to demand access; artisanal fishing is difficult to define, and fishermen have pressed for higher quotas, for the lifting of shark fishing, and for the opening of non-traditional sea urchin, octopus and squid fishing. A forceful protest by fishermen in 2004 led the government to make concessions that were only repudiated after an election.

In 2007, the United Nations Educational, Scientific and Cultural Organisation (UNESCO) placed Galapagos on its list of World Heritage Sites "In Danger", citing increasing population and poor management as the causes. A raft of measures was introduced by the government, with restrictions on immigration, bans on fishing and the introduction of a more systematic approach to inspecting incoming vessels for non-endemic species. This was enough for UNESCO to vote in 2010 to remove the archipelago from the "In Danger" list, but some observers argued that these were technical solutions, crisis management that did little to address root causes.

The conflicting interests of nature, economy and society

are often intensified within the confines of a small island, and this is something that challenges their co-dependence. The fundamental problems of the Galapagos remain an increasing population that lives on only three per cent of the land area (it is otherwise preserved and conserved for nature), weak administration, declining native species, the over-harvesting of resources, and an accelerating tourism industry only slowed briefly by global financial issues. And all this is exacerbated by a fractured relationship between a largely Ecuadorian population deemed by some to be lacking in awareness and concern for the special status of the Galapagos and an under-resourced scientific community. [20]

It seems that once the seal of isolation is broken, the unique particularities of a remote island are threatened relentlessly, transforming it – and reducing it – piece by piece into a microcosm of a larger mainland world. As a young man learning about the marvel that is evolutionary theory, I always presumed that the unique biotic community of the Galapagos Archipelago would be preserved for all time. Issues concerning the human communities on the islands, issues concerning tourism and population levels, resource exploitation and governance based in far away Lima – these were things that rarely filtered through. But now time has taught me that unique marvels, even remote islands, are subject to destructive forces of change more commonly associated with the mainland.

Elsewhere, examples of immediate and dramatic change that has been experienced on small islands when their "closed system" is breached are numerous. Almost all endemic birds in the Hawaiian group are under threat due to imported predators. Some species that have been rendered extinct by human contact include flightless birds like the dodo, the solitaire and related species of the Mascarene group (east of Madagascar), and the giant anole of Culebra Island in Puerto Rico – a tree-climbing lizard known only from a few specimens collected over eighty years ago. In New Zealand, the Stephens Island wren was wiped out by just one cat – owned by an unfortunate lighthouse keeper – even before it was formally recognised as a new species. In 1980, the Chatham Island black robin, which, like many island endemics is a slow breeder, was reduced to just one breeding pair before it was removed from a rock stack to a more stable forest environment.

Madeira has a long history of devastating fires since the early years of Portuguese colonisation, when a large section of the island was set alight and burnt for seven years. In September 2010, it was reported that forest fires had put the future of Europe's rarest seabird under threat. Zino's petrel was considered extinct until it was rediscovered in 1969. Having suffered the depredation of rats, cats, shepherds who used juvenile birds for food, and egg collectors, the fire in the island's central mountain massif killed twenty-five chicks – sixty-five per cent of the year's young birds – leaving only thirteen fledglings found alive in their underground burrows. [21]

On Guam, the decimation of its rail and Micronesian kingfisher populations means that they now only exist in captivity. Their demise is attributed to a brown tree snake, which was probably imported when it stowed away in US transport aircraft in the 1940s and was subsequently kept as a pet. Native to eastern and northern coastal Australia, Papua New Guinea and many islands in north-west Melanesia

– where it lives in equilibrium with other species – on Guam it has been a highly successful predator, seriously threatening the eighteen unique island bird species. There are reckoned to be about 8,000 of these tree snakes per square kilometre, and to achieve such numbers they have proved to be highly adaptable, eating just about any island mammal small enough for it to digest. It was reported in 2010 that a US-funded trial project was dropping radio-tagged dead poisoned mice into the jungle canopy, and the eradication programme was continued during the next two years. Officials report that there are few other species that could be tempted by the mice because the brown snake has eaten most of them!

The equivalent of that single breeding pair of Chatham Island black robins in the world of flora is the café marron tree, thought to be extinct in its one island location – Rodrigues, east of Madagascar – until a schoolboy found one old tree in 1980. Unsurprisingly described as "the rarest tree in the world", killed off by imported cattle, sheep and goats, its future has only been safeguarded by growing cuttings at Kew Gardens in London, and returning them to Rodrigues. [22]

Rats, cats, mice, rabbits, ferrets, stoats, weasels, foxes, dogs, goats, pigs, sheep, cattle, donkeys, snakes and reindeer all constitute an incomplete list of species that have been introduced by design or inadvertently, and which have contributed to the decline in island wildlife. In some cases both predators and their prey have been introduced, unwittingly enabling the problem to be ameliorated "ecologically" – by cats eating rabbits, for example. But, alas, this seems to be comparatively rare, and where it happens the problem still remains, albeit in a somewhat reduced form. A few historically significant reforestation programmes have had some success on very small islands, even if this has had little effect on the restoration of native flora.

Thus on the South Atlantic's Ascension Island, the Portuguese introduced goats during the sixteenth century, and by the mid-nineteenth century, with the added introduction of rabbits, sheep, rats and donkeys, the native flora was all but destroyed. The process continues to this day, with Texas fire ants being introduced by the US military. These ants could threaten the hatching of turtle eggs and the continued revival of seabird colonies that are only slowly recovering from the depredation of cats and rats.

In 1843, the botanist James Hooker convinced the Royal Navy, aided by Kew Gardens, to instigate a long-term tree-planting programme on Ascension. Trees were planted around the 859-metre summit of Green Mountain in an attempt to trap moisture from the warm south-east winds and create a humid cloud forest adjacent to the dry, arid, hot lava plains that make up so much of the island. It was envisaged that this would create a supply of drinking water for troops. Thirty years later, Norfolk Pine, eucalyptus, bamboo and banana trees were beginning to transform the landscape, reaching up to the summit of the island as a tropical cloud forest. Today, this forest includes ficus trees as well as ginger and guava. This transformation may appear to be a considerable improvement to the lava landscape with its indigenous vegetation ravaged by introduced herbivores. But the apparent success throws into relief some contentious ecological issues. The very limited indigenous vegetation – the fragile grasses and ferns that once defined Ascension – have continued to suffer in competition with what some see as a man-made novelty ecosystem, an unmanaged mess of invasive species.

Hooker can perhaps be forgiven for his good intent on Ascension, but not so the BBC, which, together with the UK and US military and other major communications companies, has had a major presence on the island. When the Corporation constructed a new village in the mid-1960s, it planted a Mexican thorn in order to stabilise soil. This is an aggressive pioneer in difficult terrain, which has tap roots penetrating as deep as thirty metres so making eradication extremely difficult, and which has been spread by feral donkeys eating the seed. It is estimated that unless the spread is halted, the thorn could eventually colonise up to ninety per cent of the land surface. [23]

In some islands it is possible to trace the effects of non-endemic introductions as a continuing phenomenon over a considerable time period. This is well illustrated in the case of Robinson Crusoe Island (formerly Más a Tierra) in the Juan Fernandez Archipelago, 600 kilometres west of Chile. As we saw in Chapter Two, this was Alexander Selkirk's home – and inspiration for Daniel Defoe's *Robinson Crusoe* – when he was voluntarily marooned there in 1704. But the island ecology had been altered substantially by human exploitation long before he became its sole occupant. The Spanish navigator Juan Fernández colonised it in 1574, introduced cattle, sheep and goats, and traded in sea lion oil and salt fish. Subsequently it was abandoned and then bequeathed to the Jesuits, who sought to replace the goats with pigs. Rats had swum ashore from ships or had arrived in the rowboats of provisioning parties. Cats were introduced to control the rat population, but merely multiplied along with the rats.

When the Jesuits abandoned Más a Tierra, it became a

sanctuary for deserters and continued to be a provisioning station for naval enterprises. Throughout the seventeenth century, the Spanish – who freed dogs on the island to destroy the pigs – competed with the English in the exploitation of sandalwood and chonta palm. Between 1750 and 1814, the island suffered increasing neglect as a penitentiary, and when this was abandoned, free colonists decimated the seal colonies, finished the last of the sandalwood, and used fire to clear the land indiscriminately. Rabbits were introduced, probably in the twentieth century, as were brambles, to act as "living fences" for livestock. [24]

Invasive rodents have been successfully removed from many islands worldwide, but the process of eradication sometimes has unexpected consequences. Crofters in the Outer Hebrides, Scotland, complained that when mink were eradicated because they were eating seabirds and their eggs, there was an explosion in the number of rats. And just one pair of rats, by a process of vigorous interbreeding, is capable of producing a family of many thousands within a year.

For example, far away on Pitcairn's Henderson Island, which is about forty square kilometres, there was reckoned to be a population of about 75,000 rats in 2011 when an intensive poison baiting programme was initiated. It is thought that this succeeded in killing all bar seventy-five individuals, but even with a success rate of 99.9 per cent, five years later it appeared that the population was fully restored. [25]

Even remote sub-Antarctic islands like Macquarie, south of Tasmania, have experienced the unexpected consequences of eradication programmes. Early visitors introduced horses, donkeys, pigs, goats and sheep, but the survivors were rabbits, feral cats, black cats, rats and house mice. With nearly a million pairs of royal penguins and seventeen threatened species of marine mammals and seabirds, the island is a popular stop-over for Antarctic cruise operators. In 1978, myxomatosis was used to eradicate at least ninety per cent of the rabbit population. In 2000, after twenty-five years of trapping caught nearly 2,500 cats (which were believed to be killing up to 60,000 seabirds a year), the cat problem was finally eradicated. But this led to a tenfold increase in the rabbit population, which had reached about 100,000 by 2006, and the number of rats and mice also soared. In 2006, an aerial poison-baiting programme was mapped out, claimed to be the largest rabbit, rat and mice eradication programme ever, and costing AU\$25 million. In 2011, the project was claimed a success after 300 tonnes of poisoned bait had been dropped by helicopters.

It is gratifying that at least some eradication programmes have been relatively straightforward and achieved desirable results. On Marion Island, in the Indian Ocean, five cats were introduced in 1949 to help kill mice. By 1977, there were reckoned to be about 3,500, accounting for an annual kill of birds estimated to be half a million. The last cat was killed in 1991, after a programme of hunting, gin trapping and poisoning.

A similar programme has been tried on South Georgia, in the South Atlantic's Southern Ocean, where rats and mice have had a devastating effect on seabird populations, killing off an estimated ninety-five per cent. After a "test mission" in 2011, what was claimed to be the largest rodent eradication programme ever attempted used three helicopters and 270 tonnes of bait. By mid-2015 it was reported that, after scattering millions of pellets baited with poison, rare birds were returning to nest on the island, and there was a good chance that rats had been eradicated. [26]

At about the same time on South Georgia, where Norwegian whalers introduced reindeer in the early 1900s, a major cull was undertaken with the aid of Norwegian reindeer experts. Several thousand animals had been culled by the spring of 2014 in anticipation that the island will be rendered reindeer-free.

It is apparent that on small remote islands, scientists feel free to experiment in what is in effect a natural laboratory, and to engage in activities that would be unacceptable on mainlands. For example, unlike rats, reindeer occupy a beloved niche in the British public imagination, and any attempt to eradicate them as non-indigenous – even in a national park like the Cairngorms in Scotland – would generate considerable opposition.

But islands are different, and so it is perhaps unsurprising that the problems relating to non-indigenous fauna and flora on small islands have generated their own Island Conservation organisation dedicated to direct action in eradicating invasive species. Noting that, of all species that have gone extinct since 1500, ninety per cent of birds lived on islands, as did eighty-six per cent of reptiles, ninety-five per cent of mammals and sixty-three percent of plants, the focus is exclusively on islands. Like similar government-sponsored eradication projects, these activities can be perceived as an affront to the best traditions of science, to be interfering with nature (of which mankind is a part), and to be complicit in animal cruelty in applying one human intervention to solve another. But this just seems possible on small, remote islands. [27]

Man Vs. Man

Cousteau's description of "fatal duality" in biological systems has its parallel in indigenous human island cultures and the "fatal impact" of colonising powers on native groups. On Easter Island, it is likely that the native tribes competed with one another to exploit and over-cultivate the land once covered in giant palms, before dying of smallpox or becoming slaves to tenant farmers who turned their island into an enormous sheep farm. It is estimated that of 10,000 indigenous islanders only about 100 survived a process that was nothing short of extermination.

The Guanches of the Canary Islands and the Caribs and the Arawak/Taino of the Caribbean all suffered the fate of contact with an aggressive foreign invader determined to create wealth at a cost to indigenous populations that was paid in disease, dispossession, enslavement, murder and extermination. The earliest recorded modern geno-cide took place on Tasmania where Victorian colonialism derived momentum from the belief that any indigenous population was undoubtedly inferior and primitive and almost certain to die out anyway. The last full-blooded Palawa died in 1876.

The Yaghans, canoe nomads living amongst islands around and south of the Beagle Channel in Tierra del Fuego, succumbed to disease so that as of 2017 there was just one person left alive. The Ona and Haush living further north suffered the additional intrusion of a gold rush and a boom in sheep farming. For the crime of "poaching" on their own land, they were hunted down for a bounty, a practice that continued into the early twentieth century. The last pure-blooded Ona died in 1974.

Elsewhere, the processes associated with impacting cultures has been less cataclysmic but more insidious. In human communities that are isolated and with characteristic limited resources, survival demands forms of adaptation in seeking to achieve fundamental needs. The island group of St Kilda sits some eighty kilometres off the west coast of the Isle of Skye and about 175 kilometres from mainland Scotland. The group comprises Hirta, Dun, Soay and Boreray. The population seldom exceeded 200 and was a remarkable example of particularised adaptation in a very limited natural environment.

Charles MacLean's work *St Kilda – Island on the Edge of the World*, first published more than forty years ago, remains the best general source of historical and geographical information. The islanders' exploitation of puffins, gannets and fulmars was the key to their survival, and records show they consumed 22,600 gannets in 1696 – 113 for each islander. In 1876, it was said that 89,600 puffins were killed for food. Fulmars were caught for their feathers, meat and oil, and were used to make medicine and even boot polish. With an avian dependency like this, it is perhaps unsurprising that St Kilda's history is much concerned with how its people fared as a "bird culture". [28]

It was an isolated society in which the range of personal contacts was limited and intense. Its very existence depended crucially on individuals fulfilling community tasks, in the politics of the island parliament, in the economy of fowling, fishing and crafting, and in parenting and teaching children to fulfil future island needs. In common with other remote islands in the Outer Hebrides, survival depended on nurturing shared social patterns, in "co-operative endeavour in the face of a hostile climate and a bleak terrain; fiercely strong ties of kinship, an almost mystical preoccupation with the mythical past and an immensely imaginative oral tradition of storytelling, song, verse and anecdote. People had plenty of time to rehearse their history." [29]

So the community on St Kilda was well adapted to the particularities of its limited environment. It exploited its single valuable resource efficiently and engaged in other communal subsistence and trade activities that fostered social cohesion. But such limited subsistence is precarious and fragile, not only because bird populations are subject to the vagaries of weather and food availability. Like small, isolated biotic communities, the people of St Kilda had always been vulnerable to events initiated from outside its tight-knit confines. So, when smallpox was brought to the island by a party returning from Harris in 1727, there was no immunity and – excepting a group that was absent hunting on outlying Boreray – the disease wiped out all but eighteen of the population of nearly 200. And there were disease hazards within the community, too. *Tetanus infantum* was first recorded in 1758, and by the end of the nineteenth century it was accounting for about eighty per cent of infant mortality. The probable cause was the *Clostridium tetani* that has been found in soil on the island. Possibly derived from bird products, it would have contaminated knives or scissors used for treating the umbilical cord of newborn babies. [30]

Historical events acted to expose this vulnerability across

a number of dimensions. Small islands seem to offer ideal conditions for nurturing exceptional religious beliefs, and Christian missions have performed a significant role in many parts of the world, spiritual unity contributing to a more general social cohesion and a sense of purpose to guard against isolation. In the harsh environment of St Kilda, people were probably both superstitious and believers in some form of God, something that was sustained through the myth and legend of a strong storytelling oral tradition. When an evangelical preacher visited the islands between 1822 and 1823, he described a people who had been without a resident minister for more than a hundred years as steeped in a mixture of "pagan belief and Popish superstition". This, together with their propensity for dancing to the fiddle, mouth music and various games and races on the beach of Village Bay, precipitated the arrival of an evangelical Presbyterian minister in 1829. This may have contributed to an increasingly fatalistic attitude, as successive missionaries and ministers established a kind of theocratic rule that could subdue local democratic spirit. [31]

By the end of the nineteenth century, prospects for St Kilda's minimal export economy were bleak. Mineral oil was replacing fulmar oil, dried and salted foods like fish were much less in demand, and there were alternative markets for down and feathers. Islanders began to look outside their own community for work, giving rise to concerns that there would soon be insufficient able-bodied men to crew boats visiting outlying stacks for birds.

From about 1860 until World War I, St Kilda was briefly "in vogue" as regular summer steamers deposited visitors on its shore, keen to see this "primitive society" in Victorian

Britain. Along with steamer passengers, increasing numbers of trawlers anchored offshore, and the crews were often generous in donating coal and food – charity which the St Kildans were offered from several other sources too, and which they accepted willingly since possessions meant little to them. Their self-centred independence – the product of remoteness, isolation and insulation – was replaced by increasing dependence on the mainland. The islanders, viewed as anachronistic by outsiders, became self-conscious and uncertain of their values. [32]

To live a life that anticipated the outcome of the next year as, with any luck, no worse than the one before – this was no longer acceptable. Profound changes in the nature of economic life, changes that favoured the concentration and division of labour, fed expectations of increasing growth, demand and expenditure, and commensurate growth in comfort and wealth. Just surviving and feeling at home in a place meant very little in this brave new world where "isolation was now a symptom of backwardness and isolation a kind of failure". [33]

The effects of acculturation culminated in 1930 when the islanders petitioned to be taken off the island. When a journalist visited St Kilda just before the evacuation and talked to the islanders about their situation, he found thirty-five inhabitants, of which eight were able-bodied men. The population had halved in ten years. The tasks necessary to sustain life were described as "plucking" wool from the sheep, weaving wool into tweed, the maintenance of buildings and walls, the cultivation of crops, and the perilous task of catching sea fowl. The export of feathers and fulmar oil had practically ceased, and was replaced by the sale of tweed and souvenirs to tourist boats. This provided a range of family incomes from £25 to £45 per annum.

The islanders reported that the previous winter had been exceptionally severe, and that on two occasions frenzied calls for help were transmitted by trawlers when islanders were seriously ill. A steamer, charted by the Scottish Board of Health, was unable to land anybody due to the stormy conditions, and had to turn back to the Outer Hebrides. The perils of isolation were overwhelming an aging population. Indeed, with the exception of the occasional homebound trawler managing to land a boat in the exposed bay, it was not unusual for several months to elapse without ships, mail or supplies. The journalist concludes in heroic terms: "Records prove that St Kildans have been for the past century fighting a battle against Nature. Their numbers early last century were far in excess of what they are to-day. The St Kildan race is fast disappearing, and their plea is an urgent one. It is the sad ending of a bitter, losing struggle against adversity." [34]

Chapter Four

The Economics of Vulnerability

Students of island economics and development issues hoping to use "islandness" as an analytic category distinct from "mainlandness" have struggled to make a convincing case. Constraints are broadly comparable with small countries in general, except in the case of transport issues, which weigh heavily on island economies. It appears that a unique case can only be made for an island of less than 100,000 square kilometres and with a population of less than 100,000 people. However, generalisations have a habit of showing up their limitations when applied to small islands, so it is worth exploring the detail to discover how economics colours island experience. [1]

Small Advantages

Small islands do have certain specific advantages, by far the most significant being a location, a landscape and a history that many of them possess and which can be packaged as an image that has enduring tourist appeal. "Paradise islands" abound in the minds of those marketing tourism, and the problematic economic effects of this are considered in more detail later in the chapter.

For similar reasons there is a global trade in buying and selling private islands, something that may have a small economic spinoff in nearby communities. In 2013, the Emir of Qatar purchased six of the Echinades Islands in Greek Ionia. Unsurprisingly, perhaps, he is expected to build a palace on one for his twenty-four children and three wives to enjoy. The emir is lucky, for even at the height of Greece's debt crisis, plans to sell off islands were met with a public outcry over trading off bits of the sovereign state. [2]

One has to look quite hard to find other general

economic advantages that small islands possess. They have been uniquely placed to act as bases for fishery exploitation in remote areas, most importantly before technology enabled large vessels to be independent of their services. Islands often have resilient traditional subsistence systems of production that may be sustainable in times of economic crisis. They have a few unique products to their name: orchids of Hawaii, kentia palm of Norfolk Island (although this was introduced from Lord Howe Island), coco-de-mer in Seychelles and the dragon tree of the Canary Islands. But such examples are few and far between.

Where these specific products are missing, islands have sometimes been successful in establishing "niche products" that are of high quality or "unique". The Hebridean islands of Jura, Islay and Skye have achieved this with malt whisky, as has Pitcairn with its expensive honey and related propolis ("bee glue") products.

Island societies are sometimes characterised by conservatism, something that is born of isolation and which militates against change. But one would be wrong to conclude that the consequences of this are universally negative. In Barbuda, for example, the inhabitants maintained many of what they perceived to be appropriate agricultural methods in the face of efforts by colonial administrators and rapacious proprietors who viewed the methods as primitive. To hold on to "primitive" often demands determination in the face of potentially overwhelming forces of modernisation, and it requires the acuity to demand a degree of independence that is otherwise absent on the mainland. Thus, Orkney islanders have campaigned actively against potentially lucrative uranium mining, and the community in Shetland has mitigated the negative impacts of the oil industry by exercising strict environmental controls and by lobbying for the right of the port authority to charge disturbance tax. [3]

A variety of services that islands are specially placed to offer include acting as laboratories for protecting gene pools to ensure biotic diversity – for example, Kangaroo Island in South Australia specialises in producing pathogen-free seeds. On Svalbard in Norway, the Global Seed Vault is built into an Arctic mountainside and designed to protect crop seeds against cataclysms like nuclear war. In September 2015, it was reported that researchers in the Middle East had requested samples of wheat, barley and grasses suited to dry regions to replace seeds in a war-damaged gene bank near the Syrian city of Aleppo. Long-term scientific research into the kinds of unique biotic communities that colonise remote islands is established on Aldabra, in the western Indian Ocean, and in the Galapagos. Monitoring of the atmosphere and hydrosphere for pollutants – something that cannot be carried out close to large human settlements – is performed on Mauna Kea, Hawaii. And islands situated in the path of air masses approaching continental popula-tions, like the Azores and Tristan da Cunha, act as major weather stations.

The romance and isolation of islands seems to appeal to the philatelic imagination, and the issuing of postage stamps can have a disproportionate impact on small island econo-mies as collectors buy up "rarities". On Tuvalu, the sale of stamps has exceeded AU\$1 million per annum, enough to cover the government's recurrent budget. And even on larger islands like Tonga, the sale of stamps has provided up to eight per cent of government annual revenue. Chatham, Christmas, Norfolk, Pitcairn, Easter, Ascension and the Falkland Islands, Trisdan da Cunha and St Helena have also benefitted from being "stamp islands". [4]

In Curaçao, Singapore and the Canary Islands, tax-free zones have been established to encourage specialist manufacturing industries, and elsewhere financial services are a source of foreign exchange, with so-called tax havens being set up that do not weaken the very limited domestic tax base and actually widen the otherwise narrow local economy. The Caribbean has a high concentration of islands selling financial services in exchange for favourable tax rates. The British Virgin Islands, with a population of 30,000, is reckoned to have a million offshore companies, whilst in the Cayman Islands the adult population is outnumbered by the number of registered companies. In the British Isles, Jersey, Guernsey and the Isle of Man fulfil a similar role. In some cases historically significant institutional structures specifically favour the establishment of "offshore" finance houses. Jersey, for example, is neither a British colony nor an overseas territory, nor is it officially part of the UK, and it has an international identity separate from the UK. It has been governed through a rather limited system of democracy, with no representatives in the London parliament. With very low rates of goods and services tax, as well as low income tax, and with no capital gains tax or estate tax, it operates offshore banking with a sophisticated infrastructure. It is perhaps unsurprising, then, that Jersey has a higher per capita income than Luxemburg, and a wealthy population that one suspects prefers island affairs to be not too closely inspected.

The Island

Singapore, Hong Kong, Vanuatu, Nauru, the Cook Islands, Mauritius and Bahrain are among other islands and island states that administer various forms of low-tax or no-tax institutions where security and discretion can be guaranteed. However, those who have campaigned for a fairer tax policy that does not favour society's richest will be heartened that guarantees of confidentiality are being increasingly investigated by fraud squads and government policy-makers alike. At the beginning of 2013, in what was seen as a concerted attack on tax havens by the United States and the European Union, the British government began difficult negotiations to reveal details of people using financial institutions in its crown dependencies and overseas territories, including the British Virgin Islands, Cayman Islands, Anguilla, Bermuda, Montserrat and the Turks and Caicos. Whilst the British government anticipated recouping a billion pounds in this process, such good housekeeping by the mother country bodes badly for the economies of her charges. [5]

Under the United Nations Convention on the Law of the Sea (UNCLOS), dependent territories and nation states have been able to extend their economic jurisdictions very considerably beyond their twelve-nautical-mile territorial limit. Exclusive Economic Zones (EEZs) provide almost complete rights to the living and non-living resources of the seabed, and substantial rights to the water content. In order to generate an EEZ of 200 nautical miles (370 kilometres), an island needs to be permanently inhabited and engaged in economic activity. Distance from competing islands is at least as important as size: the Convention means that an inhabited isolated island of one square kilometre can generate an EEZ of 125,664 square nautical miles (431,015 square kilometres). The effect of applying this ratio to the remotest islands is dramatic, and some of the more remarkable examples are included in the table below. [6]

	Land area (sq.km.)	EEZ (sq.km.)
Pitcairn	5	800,000
Tokelau	10	290,000
Nauru	21	320,000
Tuvalu	26	900,000
Norfolk	36	400,000

In contrast to the huge sea areas generated by these extremely remote islands and island groups that are capable of sustaining habitation or an economic life, a smaller island – commonly termed a "rock" in matters of jurisprudence – will generate only the twelve-nautical-mile territorial limit, amounting to 452 square nautical miles. Some nations with huge landmasses but with relatively few far-offshore islands, or where the islands compete with other nations for jurisdiction, generate small EEZs. Canada, with its huge bays, gulfs and island-studded Arctic Ocean, has a zone that is actually smaller than its territorial limit. France, in comparison, with its overseas territories in remote archipelagos, can claim an EEZ nearly twenty times the size of its landmass.

The arithmetical challenges of EEZ designation are particularly complex. For example, China has claimed 1.4 million square nautical miles based on about 250 rocks, reefs and islands that it has enlarged significantly by dredging sand from surrounding shallows and which collectively add up to about six square miles. And even if China's claims were accepted by UNCLOS, the International Court of

Justice will not re-write geography on account of a tiny island. (For example, Colombia's Serrano and Quita Sueño Islands, which are adjacent to a lengthy section of Nicaragua's coast, were enclaved in a twelve-nautical-mile territorial sea surrounded by a Nicaraguan EEZ.)

Nearer to home, when the United Kingdom ratified the Law of the Sea in 1997, the government agreed to use inhabited St Kilda as a point of reference rather than uninhabited Rockall, thus conforming to the regulations. In the process, the UK gave up claims to nearly 100,000 square kilometres of sea compared to the previous zone calculated from Rockall. However, the UK continues to claim tiny Rockall, which was the object of Britain's last imperial adventure when it annexed the rock in 1955. This is not something that sits happily with Ireland, Iceland and the Faroe Islands, which have long-established claims of their own on this islet, which is some 400 kilometres from the Scottish mainland. In fact, even using St Kilda as the reference point may be open to challenge. Its supposedly "permanent" population could quite fairly be described as "transient", comprised as it is of a few workers at the missile-tracking station, and those who work for the National Trust for Scotland, which owns the island. And its supposedly active economy could equally fairly be described as so small as to be non-existent since the people were evacuated in 1930.

The advantages of these zones for island economies has yet to be fully realised, but benefits could end up being more apparent than real. For example, a small island nation state in the Pacific will still have to compete with large colonial powers like the United States and France, who will also gain extensive EEZs. The power to enforce economic law against huge industrial interests like the Japanese and US fishing fleets in the Pacific is, in reality, limited; and in general the financial resources and sophisticated technology necessary to search for and exploit oil, gas and minerals is unavailable and would call for overseas aid that could merely reinforce foreign dependence.

The serendipity of island location means too that economic advantage will be very variable. Where several island states are proximate and close to a powerful coastal state, the competition for marine resources could actually be more fierce as countries strive to realise jurisdiction over space which the Law of the Sea has only relatively recently served to confirm. For example, Caribbean coastal states like Colombia and especially Venezuela have benefitted from extended jurisdictions that have enabled them to absorb the traditional fishing areas of other nations.

Whilst French Polynesia could generate an EEZ of one million square kilometres, this compares with the island states of the eastern Caribbean in particular which, because they are small, numerous and proximate to each other, will gain very little jurisdiction. [7]

There is a paradox here in that whilst remote island states can inherit enormous EEZs, they inherit too huge challenges in ensuring that exploitation truly brings long-term benefits. Alas, as we will see, strategic planning has been little more than a pipe dream in many instances.

Finally, islands have certain strategic advantages achieved through their remoteness. This enables them to be perceived as appropriate locations – often regardless of the wishes of indigenous islanders – for testing weapons of mass destruction, storing radioactive waste, military bases and landing grounds, telecommunication and satellite stations, the bulk-handling, transhipment and treatment of dangerous substances, exile, imprisonment, quarantine stations, leprosariums ... locations suited well to the dystopic imagination.[8] We will look further at the significance of strategic location in Chapter Six, particularly with regard to tourism, militarisation, and using islands to isolate human activities that would rather not be seen on the mainland.

Natural Disadvantages

Any listing of the meagre economic resources islands are able to draw upon paints an incomplete and partial picture. It would be wrong to make many assumptions about the social health of island communities in general from a picture of economic limitations. For example, E.C. Dommen sampled twenty-four islands and compared them with "continental counterparts" with similar GDP. Across a range of geographical, social and economic factors, his somewhat quirky findings suggest that on islands: earthquakes are less common and less damaging; hurricanes are more common; the environment is probably more bounteous; biodiversity is limited and vulnerable; islanders probably live longer; population is more dispersed – but not less urbanised – and is growing more slowly and with lower crude rates of birth and death; emigration is a normal feature; school enrolment is relatively high; and internal disorder is less common in spite of the fact that islanders appear more willing to tolerate separatist movements. Dommen concludes that islands are "particularly fortunate places, where life is longer and nature is bounteous though the menu may be short. Politics are friendlier. Hurricanes are more dangerous than social unrest." [9]

On the more significant debit side of economic standing, the poor status of islands is illustrated in the fact that the 2011 Gross Domestic Product per capita ranking of 193 states places eighteen small island states within the bottom twenty. Gross Domestic Income per capita ranking for the same year paints a somewhat less bleak outlook, and of the 215 states ranked, sixteen of the forty small island nations are in the top half of the rankings. [10]

The disadvantages that small islands suffer tend to be expressed in classical economic and geographical explanations. Disadvantage is characterised by a narrow range of resources fostering 1) precarious specialisation and the prospect of premature depletion; 2) economic dependence on larger countries for markets, investment, and very significantly transport; small populations, high migration, a limited pool of skills making it difficult to generate economies of scale and generate specialist expertise; high population densities making high demands on resources; and limited space, with centres of production in vulnerable coastal locations.

There tend to be intimate linkages with ecosystems that increase the chance of entire islands being made unproductive or even uninhabitable unless costly impact controls are put in place, and there is a high ratio of coastline to land area that leaves islands highly vulnerable to a range of extreme marine and climate influences especially related to global climate change. As we've already seen, islands are more disaster-prone, and isolated islands often have specialised and highly adapted ecosystems that are vulnerable to non-indigenous species introduction.

Islands are vulnerable to the vicissitudes of commodity prices, which leaves producers with little influence over terms of trade and can lead island nation GDP to fall by up to thirty per cent in one year. There is a perception of island economies as high-risk entities by global investment agencies; the globalisation of free trade systems discourages older preferential market agreements traditionally offered (for example, by their ex-colonial governments); and there is sometimes dependence on international aid and migrant remittances, which are both insecure. [11]

There are exceptions to the rule that small island nations have limited natural resources, and growth can be focused on exploiting something that is in world demand. For example, Trinidad (oil), Jamaica (bauxite) and New Caledonia (most importantly nickel) are well endowed. But their economies are vulnerable to the day these resources are exhausted, which is the case with Fiji (gold), Vanuatu (manganese), and Banaba, Makatea and Nauru (phosphates).

Economic Ups and Downs
Specific island experiences illustrate some of the economic generalities we've been considering. Ocean Island (now Banaba) is six square kilometres in extent, and situated about 4,500 kilometres north-east of the Queensland coast of Australia. It was annexed in 1901 by the British government soon after the Pacific Island Phosphate Company bought a 999-year sole mining right for £50 a year. Composed almost entirely of a thick layer of phosphate derived from guano – which means "dung" in Spanish, and is material formed from many hundreds of years of seabird and sometimes seal droppings – intensive

strip mining was operated by what became the British Phosphate Commission (BPC). Phosphates are used throughout the world where modern intensive agriculture is practised, and it is therefore depressingly ironic that on Ocean Island surface mining rendered useless huge tracts of land whilst islanders actually found themselves deprived of even pockets of fertile ground. Small agricultural holdings have traditionally provided a vital element in island subsistence economies and in the social infrastructure of many Pacific islands. So whenever the people of Ocean Island mounted a concerted protest, the company increased the extraction payments. But it is claimed that successive British Resident Commissioners coerced the population into accepting low rates of reimbursement: £35 to £50 for phosphate-rich land that was worth up to £2,500 per hectare to the BPC. [12]

By 1941–42, the British government began to realise their commercial interest was rendering anything else unsustainable. Their solution was not to stop mining, but instead – in an act of grossly misplaced paternalism in keeping with imperial traditions – to seek a new home for the islanders on their behalf.

Early in the second half of the nineteenth century, a Fijian chief from Taveuni sold Rabi Island, nearly 2,500 kilometres from Ocean Island, to Lever Brothers for use as a copra plantation. By the early stages of World War II, they were willing to sell this to the British government for £25,000, money that was taken from the inhabitants of Ocean Island's phosphate royalties. But any chance the islanders had of inspecting plans for their new home was cut short by the Japanese invasion of Ocean Island. Many people were murdered and others suffered gravely in internment camps on surrounding islands.

At the end of the war the BPC and British authorities informed the remaining islanders that the Japanese occupation had rendered Ocean Island uninhabitable for an agriculture-based population. This was not true. Some 1,002 islanders were relocated on Rabi, and the BPC continued mining. Others followed in 1977 and still more between 1981 and 1983, although some returned after mining ceased in 1979. By this time some ninety per cent of the island surface had been laid waste.

Photographs of the Rabi "camp" support the assertion that there were no houses to move into as promised by the British government. There were about three months of canned rations, it was the hurricane season, and many of the people were still weak from years in Japanese internment camps. They were unfamiliar with shallow reef fishing and unskilled in agricultural practices. One wonders how the descendants of the displaced inhabitants of Rabi, now living on nearby Taveuni, may have felt as they helped the arrivals through this difficult period. Reports concerning hardships experienced by the people differ, but there is no doubt that there was a painful "adjustment period" and that very soon there was a move to return to Ocean Island. Compensation needed to achieve this could only be realised after more than ten years of proceedings through British courts. In 1976, AU$10 million was set aside for development and community work, but a stipulation demanded it be used on Rabi only. [13]

There are currently only about 300 people living on Banaba (as Ocean Island was renamed when it became part of the Republic of Kiribati in 1979) and about 5,000 Banabans living on Rabi.

Another so-called "rock island", Makatea, in the Tuamotu Group of French Polynesia, suffered the same fate of evacuation when phosphate deposits were exhausted in 1966. The twenty-four-square-kilometre island currently has a population of about sixty.

The history of the island state of Nauru follows a more prolonged trail of disasters. Situated some 4,250 kilometres from the Queensland coast of Australia, it has a population of about 10,000 and occupies twenty-six square kilometres, of which eighty per cent is uninhabitable due to strip mining. After a century of extracting high-grade guano, the interior has been left a maze of five-metre-high jagged stumps – all that remains of the original coral limestone bedrock.

A whaling ship was the first recorded Western visitor in 1798. Captain John Fearn found an "extremely populous" speck of land, which he named Pleasant Island. Its remoteness, along with an absence of nearby islands and a westerly flowing equatorial current, had all combined to discourage exploratory travel in the area. The island seems to have been well endowed with food resources: coconuts, mangos, breadfruit, pineapples, pandanus and wild almond trees grew well, and its coral reefs were well stocked with fish. It seems to have been a self-contained society that was largely ignored by Western explorers who, for the moment, could see no significant natural resources.

During the next hundred years Nauru suffered, like so many Pacific islands, from a panoply of interrelated impacts that involved tribal warfare, colonialism, Christianity, immigration and disease, but it was to be the mining of phosphate that would dominate the island's lifeworld.

No island, no matter how small, remote and seemingly isolated from mainland interests, could remain secure from the dynamic and opportunistic forces of colonial capitalism. In 1886, Germany was granted dominion over Nauru as part of an agreement with Britain establishing each country's dominion in the western Pacific. By this time, Germany had already discovered Nauru to be a fertile island that could be exploited for the copra trade. And when a few years later it was discovered that Nauru was endowed with vast deposits of some of the richest phosphates in the world, its fate was sealed in the most ironic way. Described as "scores of millions of tons which would make the desert bloom as a rose, would enable hard-working farmers to make a living, and would facilitate ... [food production] ... for hungry millions for the next hundred years to come", what was not divulged was that during this "next hundred years" Nauru's environment, culture and society would be torn asunder. [14]

While the Nauruans were not part of any formal agreement, the Germans paid the islanders a very modest amount per ton of rock removed from their land. In the aftermath of World War I, Britain, Australia and New Zealand were all granted access to Nauru's precious resource. With Australia fulfilling a dominant role, the British Phosphate Commission offered a small phosphate royalty, regarded as fitting for a people who lived on "coconuts, fish and sunshine".

The effects of this benevolence were swept away during World War II when the island was captured by the Japanese. Many Nauruans were transported to forced labour camps on the island of Truk, from which only 737 emerged. In the post-war recovery, Australia tried to do what the British administration had done on Banaba, and claimed that it would cost them hundreds of millions of dollars to rehabilitate the islanders. The Nauruans were offered Curtis island, off the north Queensland coast, and seem to have avoided being "persuaded" to go there only because they were recognised as a readily available labour force that would ensure the supply of the fertiliser upon which Australia's meat industry in particular was dependent.

Phosphate continued to be extracted at an exponential rate, but a still negligible fraction of the phosphate revenues were given to the Nauruans, and it was not until 1966 that domestic and international pressure moved the Commissioners to give up larger percentages. The Nauruans nevertheless demanded control over their island, and after a bitter struggle, Nauru was finally granted independence on January 31, 1968. However, as with many post-colonial states, Nauru did not enter the world system on an equal footing.

Sixty years of mining had radically disrupted the nation's ecosystem and eroded the land – meaning that the Nauruan way of life and the intricate relationship with their surroundings was also eroded. After thousands of years of self-sufficiency, colonialism had acted to transform the indigenous belief and value system and its culture. An integral element in the Nauruan lifestyle was its complete dependence on the tiny island, which the islanders used for both their livelihood and their enjoyment. Now they bore the brunt of dealing with seemingly irreparable environmental damage and a land shortage due to mining. [15]

In 1989, Nauru initiated proceedings against Australia at the International Court of Justice (ICJ) on the basis that it had violated several of its international legal obligations. On August 10, 1993, a settlement was reached and Australia agreed to award AU$107 million to Nauru over twenty years as compensation for environmental damage. This seems like a significant sum, but less so if one considers that, in 1967 alone, mining generated about $300 million. A compensation figure of just five per cent compounded for each year between 1967 and 1993 – when the island was under the trusteeship of Australia, New Zealand and the UK – would have exceeded a billion dollars. [16]

Nauru became one of the smallest republics in the world, but the richest island in the Pacific, with a per capita income exceeding that of even the United States and Saudi Arabia. But the islanders had long been aware that their prosperity was based on phosphate deposits that were more likely to be depleted soon after the end of the century, rather than an earlier optimistic estimate that gave mining another 200 years of life. After independence, they concentrated on rebuilding population and in assembling a trust fund to sustain them in the post-phosphate future. Even in 1993 it was reported that "the property they now own includes urban land in Australia, the biggest hotel in Guam, a big commercial and residential block in Hawaii and fertiliser plants in India and the Philippines." It was reckoned that they had sufficient financial resources to ensure that, *if properly used*, a comfortable long-term livelihood could be guaranteed. [17]

Even at that time, the trust fund was instructing lawyers to press a London-based solicitor on what had happened to $60 million of its money. But lifestyles were lavish, jobs were plentiful, housing free and no one paid tax. Children went to good schools in Australia. There was an airline with seven aircraft and a shipping line with several ships.

Things declined so quickly in this new island state that it soon became difficult to believe one was reading about the same country that so recently had been famed for its wealth. In 2004, not much more than ten years since it had an estimated $700 million in trust, the government of Nauru had no money and no assets, a fact driven home by a default on a $236 million loan, forcing an American financier to seize all the nation's property assets worldwide. In 2008, the debt crisis was reported to have reached AU$1 billion. Unemployment was running at about ninety per cent. Kathy Marks, in an article titled "South Pacific Tragedy", described "rows of concrete shacks on the Nauru foreshore [that] look derelict, with their cracked louvred windows, smashed walls laden with graffiti, and narrow passageways strewn with junk. But many of these miserable dwellings are inhabited – by the poorest of the poor." There now seemed little chance that the eagerly anticipated tourism industry could develop with workers going unpaid for months, a barely working telephone system, the "airline" now a single plane (which was once even wheel-clamped), and an abandoned and overgrown hotel, swimming pool and golf course. [18]

Victim of compound swindling by corrupt overseas investment houses, by consultants of dubious repute and by their own leaders, victims of mismanagement and the sheer profligacy of islanders, one wonders what Nauru's future would have been had its government shown wise leadership.

It is not as if they did not try to develop new initiatives. In the early 2000s, they promoted a zero-tax regime to attract offshore banking. The government charged £25,000 to set up a bank registered in Nauru, and over 400 banks were "established" – really just post box addresses with no physical presence on the island. By mid-2001, Nauru was under scrutiny by the Financial Action Task Force investigating allegations that billions of dollars of Russian mafia money was being laundered through its banks. [19]

At the same time, the government accepted an approach from Australia to establish asylum seeker centres on the island. These "boat people", mostly from Afghanistan, Sri Lanka, Iraq, Syria and Iran, were trying to reach Australia's Christmas Island via Indonesia. Millions of dollars were spent to set up a detention camp where would-be refugees intercepted at sea were incarcerated under the notorious "Pacific Solution". Australia set up another centre on the island of Manus, in Papua New Guinea, during the mid-2000s.

The policy may have been condemned internationally as a serious contravention of human rights, but it has provided vital capital for Nauru. Australia resourced substantial infrastructure, thousands of jobs were provided for locals, the hospital and schools were refurbished, and power and water supplies were upgraded. Then, at the end of 2007, there was a change of government in Australia and the project was scrapped. Desert mining camps were prospected as an alternative location for internment, but not for long. In August 2012, a new government reinstated the policy and reopened both Nauru and Manus in response to increasing numbers of refugees approaching the coast of Australia, and in particular in response to a tragedy in June 2012 when two boats capsized north of Christmas Island, killing at least ninety people. (In March 2011, Christmas Island – an Australian territory about 2,500 kilometres north-north-west of Perth, and the site of the country's largest asylum seeker detention centre for people who had arrived in Australia "conventionally" – was subject to several days of violent rioting and demonstrations against delays in processing applications.)

The re-revised policy was reinforced in 2013 with an announcement that "boat people" whose application for asylum was successful would not be resettled in Australia, but in impoverished Papua New Guinea; and by September 2014 it was being reported that the government had reached a "secret agreement" to send migrants to Cambodia – a country with no history of accepting refugees, that was mired in human rights abuses, and that had been swayed by a £22 million "financial incentive". Migrants probably will not go to Cambodia, but the need to find somewhere else for them to be detained has only been made more pressing by, in May 2016, a Papua New Guinea court ruling that their detention centre on Manus was unconstitutional. Worse still, in early 2017 Nauru was headline news when newly elected US president Donald Trump telephoned Australia's prime minister Malcolm Turnbull and described the previous administration's commitment to resettle up to 1,250 asylum seekers in the United States from Nauru and Manus Island as "a dumb deal".

The "Pacific Solution" reflects an attempt on both sides of the Australian political spectrum to present a hardline stance on an issue that voters in marginal seats have cited as one of their chief concerns. The government's rationale for using Nauru and Manus reflects a belief amongst some Australians that no advantage should be gained by even genuine asylum seekers undertaking a dangerous voyage by circumventing regular migration arrangements. People who do this constitute only about two per cent of total migration, and the majority have been found to have a valid case for entry with large numbers spending years behind razor wire. Described by the Refugee Council as "a failed policy that created great psychological damage and undermined Australia's reputation on human rights", the policy was branded a scandal internationally and among many Australians. An unknown number of asylum seekers have drowned, perhaps between 500 and 1,000. This is a shocking indictment, but it is made even worse by an Amnesty International report that claims that in May and July 2015 officials paid people-smugglers to turn back boats. In the first case, passengers were allegedly transferred to two boats, one of which sank on an island near the coast of Indonesia; and in the second reported case, the migrant boat ran out of fuel and was picked up by Indonesian officials. [20]

It is perhaps unsurprising then that whilst 300 boats carrying illegal immigrants reached Australia in 2013, by 2014 this number was reduced to one. This was bad news for Nauru, but it was already working on another project, radical even by the standards of its rather surreal history of economic and geopolitical endeavours. In late 2009, Nauru became only the fourth country, after Russia, Nicaragua and Venezuela, to establish diplomatic relations with Abkhazia and South Ossetia on the east coast of the Black Sea, countries viewed by the rest of the world as renegade provinces of Georgia. Since the end of its 2008 war with Georgia, Russia has been lobbying for international recognition of these two states, which were at the heart of that conflict. In return for this diplomatic move it was reported that Nauru received considerable assistance with work on its port and aircraft. The ultimate goal for Russia is to get Abkhazia and South Ossetia recognised in the United Nations, and Nauru will now contribute to this with its vote. [21]

Pacific island states, no matter how far away they are from global power bases, are relatively easy to persuade in their allegiances because of their small population and the poor state of their economies. It is for this reason too that both Russia and Georgia have been wooing Fiji since it was cast out of the Pacific Forum regional bloc for failing to hold elections after a military coup in 2006. Whilst Australia will react warily to Russia pushing into its traditional zone of influence, Russia responds to criticism of this "cheque-book diplomacy" by claiming that it is merely responding to the reality of global economic growth and, like all major powers, focusing more attention on Asia-Pacific. Ultimately the welfare of the people of Nauru could depend on the crumbs cast by the major economic powers jostling for position in a new world order, something one could scarcely regard as stable or sustainable in even the short term.

The story of Nauru – the dramatic, tragic and seemingly surreal consequences of phosphate exploitation – gives rise to a certain media attention that would otherwise be elsewhere. The state has brought about, or been the victim of, the loss of a huge sum of money that could otherwise have contributed to a secure future; and, now desperate for revenue, it has sold its passports, its satellite and fishing rights, and its banking licences in sometimes dubious legal circumstances and with an apparent disregard for principles of sustainability. It has allowed Australia to dump its asylum seekers on its shores, and has entertained the idea of allowing it to dump its nuclear waste there, too. It has sought the patronage of faraway superpowers and in so doing it has supported causes unpopular in the international community. Its domestic politics are plagued by instability, just as its people are plagued by high levels of obesity, diabetes and heart disease due to the prevalence of poor diets.

Whilst the Western media tends to delight in implying that much of this tragedy is self-inflicted, it has little interest in examining its causes, imbedded as they are in the island's colonial history. Nauru can be seen to represent the tragic outcome of a one-sided conflict between the forces of cultural preservation and economic development. And it is an outcome that has ominous implications not only for small, remote, post-colonial island states, but for the greater mainland world, too.

The land has been transformed into an uninhabitable, mostly dusty, arid, barren wasteland which, with its alien landscape of pillars and pits now devoid of vegetation, has helped create a microclimate so that the interior is now very hot and there is an increase in drought conditions.

This tiny island state was clearly unable to deal with how best to ensure a secure and happy future in the aftermath of its compensation agreement with Australia. Like so many people who win a large sum of money and who defy all rational sense in their apparent determination to squander it away lest it burns their hands, it is perhaps unsurprising that Nauru's wealthy status was short-lived.

What may appear less understandable, in view of the island's recent history, is why the government allowed mining operations to continue at all after independence. However, with the body of the island just a skeleton of its former self, it seems there were few alternatives. The scale of the phosphate industry was overwhelming when compared with the traditional subsistence economy. The world market devoured Nauru and its culture and everything that had, for generations, made for a meaningful life. And it presented Nauruans with the opportunity to fast-track into a Western value system based on commodity accumulation. Independence was really a Faustian pact, for the island remained dependent on the market, and the market was controlled by entrenched power structures within the ex-colonising nations that had granted independence. After decades of exploitative colonial practices, independent Nauru was really a relic of its past, and utilitarian in its need to convert a resource into a commodity.

Capitalism, with its global market, sets a negligible price on preserving biodiversity, culture and society. On a tiny, remote island the effects of this can be both compressed and magnified, unique in itself but also a microcosm of the mainland world and a warning to it of what the future may hold. [22]

A Tourist Paradise?

Nauru's wealth until the early 2000s depended on the management of a single resource – phosphate – and the investment of the capital realised by selling fertiliser. This kind of extremely limited economic base makes small islands in general vulnerable, and this is particularly apparent too in relation to the tourist industry, which operates in the face of many unstable variables. Especially in the Pacific and Caribbean, islands have been able to develop a tourist industry by exploiting an image of paradise, which has been a persistent metaphor since the early days of European intervention. Sun and sand, tranquillity, exoticism and a frisson of adventure – this is an image that sells. Where it is hard to sustain, imaginative entrepreneurs have more recently been exploiting unique features of island geography to develop the ecotourism of climate change, and adventure tourism based on observing tectonic activity "as it happens".

Developing island states have responded to the siren call of tourism, believing that it can shift their economies away from small-scale traditional activities, like farming and fishing, and away from activities with a finite resource base, like mining. But tourism brings its own set of uncertainties and dependencies. Not only is it dependent on market forces to ensure people have funds to engage in tourism pursuits – and such luxuries are usually amongst the first to be curtailed during times of economic hardship – but it is vulnerable to equally unpredictable variables like fashion, weather, exchange rate and transport costs.

In about half of the eighteen island developing nations with populations under one million for which information is available, gross receipts from tourism are larger than all visible exports put together. In some places, like the Maldives, tourism directly and indirectly generates as much as seventy per cent of national income – its vulnerability was illustrated in September 2013 when a majority of hotel workers threatened to strike indefinitely over election delays. The Bahamas too receives some seventy per cent of income from tourism, and the proportion is well over fifty per cent in Antigua, Barbados, Grenada, St Kitts and Nevis, and the Dominican Republic. [23]

However, like so many things imported into a small and remote society from far away, the effects can be magnified out of mainland-accepted proportions, so the ecological, cultural and social costs can be high. Much of the capital generated from economic activities like tourism is channelled back to overseas investors. The rest accumulates differentially within a community and goes especially to entrepreneurs who may well live much of their lives off-island. In all this, the island can seem like a stage upon which transactions are performed with the islanders, unless they are lucky, passive bystanders. This contributes to a disparity in wealth and opportunities where some level of equality was once implicit in an "underdeveloped" economy.

In a wide-ranging critique of the costs of tourism in the Caribbean, Polly Pattullo describes a number of dangers that island communities have encountered. In particular, she identifies the gap between small groups of relatively wealthy tourists and large marginalised minorities contributing to the widening of social divisions and fostering a legacy of colonialism. These problems are exacerbated as the industry has become increasingly based on the walled-off all-inclusive resort that imports everything from the United States. This means that there are poor linkages between tourism and other parts of the economy, where for example, local agriculture and fisheries could supply resorts. Indeed, successful tourist islands like the Bahamas and Bermuda import very large proportions of their food. Tourism has thus become, like the old plantation system, an industry forced to organise itself on other people's terms. Thus, some ninety per cent of the tourism industry of St Lucia is owned by foreigners, and with notable exceptions like Jamaica and Barbados, this is a pattern repeated elsewhere.

Pattullo describes villagers in a remote settlement on the rugged north coast of Dominica "surrounded by banana gardens, rainforest and the Atlantic ... [who] rarely see a tourist. Even the adventurous ones do not penetrate that far along the pot-holed access road. So the people of Vielle Case are not, for the moment, waiters, hair-braiders, or taxi-drivers; they do not sell duty-free Colombian emeralds or T-shirts printed with Vielle Case Jammin'; their young men do not sell drugs or go with young white women. Yet even these farmers and fishermen are gearing themselves up for tourism. ... Every government in the Caribbean has identi-fied tourism as the region's 'engine of growth'." [24]

It is not surprising that Dominica sees tourism as an alternative to an increasingly precarious dependence on bananas, which, since the 1960s, have been grown by peasant smallholders in what is widely regarded as a success story. Banana cultivation has enabled local people to grow and market their own produce on a small scale, and unlike other export commodities like sugar, cacao, coffee and coconuts, the proportion that was not exported could be eaten. However, in reality, the export and marketing of bananas in the Windward Islands has been monopolised by a single multinational corporation – Geest, and its succes-sors – who also set the price. So the farmers have to bear all the risks of cultivation without gaining any freedom of action at the market, and the giant's share of the profit is enjoyed by investors outside of Dominica. And they can expect the same if they "diversify" into tourism. [25]

Those farmers and fisherman in rural Dominica anticipating the arrival of the tourist industry might do well to avail themselves of what has become known as the Butler Model of tourism's life cycle. According to this model, development is initiated by the discovery of an unspoiled location. There follows an intensive period of high investment as the infrastructure is built and mass tourism arrives. In time, the investment potential declines and the big spenders move on. The enterprise begins to stagnate, requiring rejuvenation that necessitates more investment – the sources of which have departed – or low-cost modification in the form of down-market tourism. If this fails then the project is asset-stripped and abandoned. What was once poor and unspoiled is again poor but now spoiled. [26]

Tourism provides an infrastructure in which organised crime, such as that involved with the trade of drugs, can flourish. The Caribbean in particular, with its proximity to source areas in South America, has a network of airline and cruise ship connections, a large and mobile volume of tourists, easy navigation, and inadequately patrolled waters where policing amongst small island mini-states is inherently difficult. In 1989, the British publication *The Economist* estimated that about $25 billion in drug money had been laundered through the tiny Cayman Islands in that year alone. Institutions of some Caribbean countries have been so effectively penetrated by vast amounts of money generated by organised crime that, according to a former head of the Commonwealth Secretariat's Commercial Crime Unit, they may justly be described as criminal states. It has been suggested that, until recently at least, drug money has been a major driver of the real estate and gambling industries in Aruba. In a 1994 US Bureau of International Narcotics Matters report, almost nowhere in the eastern Caribbean escaped censure, with Barbados, Antigua, St Vincent, the Grenadines and St Kitts considered island groups where senior officials allow drug activities to continue regardless of official policy. Curaçao, St Maarten and the Turks and Caicos – all places where tourism and financial services industries come together – also have malodorous reputations in this respect. [27]

The Turks and Caicos deserve particular attention. Twenty-eight islets covering about 250 square kilometres, this British overseas territory is home to some 30,000 people who accommodate some 260,000 tourists each year. Crime, and the political corruption it attracts, is best kept out of international public knowledge when governments wish to promote tourism. In 1985, Chief Minister Norman Saunders, together with two government members, was arrested in Miami by FBI agents. Saunders was accused of bribery and of permitting safe passage of drugs by allowing stop-over refuelling during flights from Colombia to the United States. He subsequently spent eight years in prison, but such was the close relationship between crime and politics that he was available to contest a seat in the 1995 election.

Jump forward a few years, and Premier Michael Misick resigned in 2009 after a British parliamentary committee charged that he had built up a multi-billion-dollar fortune since being elected in 2003.

Only then did the British Foreign Office in effect suspend parliamentary democracy, and appoint a governor in its place. He found himself presiding over an economy that had benefitted from a decade-long development boom making the Turks and Caicos one of the fastest-growing island groups in the Caribbean, but a boom that collapsed with the global recession. One of the locally owned banks closed its doors too, wiping out the savings of thousands of depositors and businesses. The governor found debts of tens of millions of dollars, with the government unable to pay its bills and trying to impose swingeing cuts. Many islanders demanded a financial bailout from the UK for allowing this situation to develop, but there appears to have been very little support for suspending the constitution with the prospect of tight control and cuts in the budget. Local politicians soon began calling for elections, with partisans pointing out that in this small community business had always worked "pragmatically" through personal relation-ships, where party affiliations were often dictated by family, and where the exchange of gifts was customary.

What was, in effect, direct rule ended in 2012 with the election of Mr Misick's former party, the Progressive National Party under new leadership; but it was not until mid-2015 that seven Supreme Court judges rejected a challenge from Misick to allow a corruption trial to go ahead without jury against the former Premier and several other members of his government. A fair trial by jury was considered impossible on a tiny island where potential jurors would have connections with the defendants. [28]

The corrupt activities in the Turks and Caicos have been well publicised, mainly because the islands are a British overseas territory with a large "home audience" following events. This is less so in a place like the Providence and San Andrés Islands, 220 kilometres east of the Nicaraguan Atlantic coast. Settled by English-speaking London Puritans in 1631, their significance now lies in the fact that they can generate about 350,000 square kilometres of territorial waters. But that was for the future, and in the meantime they attracted no formal British presence, and had been considered part of Colombia since 1830. Indeed, they received little international attention until, in 1953, the Colombian military dictator General Pinilla described the 250,000 English-speaking indigenous people as "undependable" and "genetically predisposed to eroticism". He initiated a policy involving high rates of immigration and daily flights from Colombia, declared the islands a tax-free zone, and invested heavily in all-inclusive resorts. Tourists were thus able to take advantage of a tiny, tax-free retail world.

Unsurprisingly, the significant English-speaking minority feel discouraged by the events of the last fifty years. This is especially true since there are now over 100,000 people living on a thirteen-kilometre strip of land, where many of the immigrants are illegal, and escaping civil war and economic recession at home, and where crime rates are high enough for the islands to have gained the reputation for having the highest kidnap rate in the world. Virginia Archbold, the island historian – reflecting the fact that, despite recent events, many people on the island apparently still feel that Britain is their mother country – claimed that "Britain should have more interest in us. We are only on this island because you sent our ancestors here." [29] Alas, I suspect Ms Archbold received short shrift from the Foreign Office in London.

The Economics of Vulnerability

In the meantime, Nicaragua has become increasingly agitated, and perhaps envious, of these activities on an island group much closer to its shores than Colombia's. It is for this reason that Providence and San Andrés was the subject of a lawsuit before the International Court of Justice in The Hague, which in 2012 ruled that the islands belong to Colombia, whilst at the same time granting Nicaragua control of a large area of surrounding seabed that could hold oil reserves.

Tourism is attracted to the tax-free perks of Providence and San Andrés, but more importantly, and as is the case throughout the Caribbean and the Pacific, tourists are attracted to a particular image. And, regardless of conflicting realities, this image is promoted energetically by the industry so that it is firmly embedded in the consciousness of potential clients. San Andrés has been a focus for smuggling for over a century, and now it is a key redistribution point for drugs travelling from Colombia northwards. Inevitably, drug money has contributed to the island economy but at a price that has included a high level of violent crime. This has not deterred its official tourist website from describing it as a "miniature paradise". [30]

Whatever the brutality of its history, whatever the indigenous extermination and slavery, whatever the havoc wreaked by hurricane and volcanic eruption, whatever the chronically partial experience of the visitor restricted to the airport, the beach, and highly controlled activities and visits to "sights", and whatever the one-sided relationship with local people who above all else must be perceived as "friendly" – whatever the enormity of evidence to the contrary, the "destination" has to be *paradise*. Tourists anticipate *their* islands as a string of pearls surrounded by the white beaches of a perfectly blue, calm ocean, palm trees, secluded coves sheltering thatched huts, which are all that is necessary in the never-ending days of warm sunshine and light scented breezes. And they must not be disappointed. The land is fruitful, and although the image portrays few local people, there is every prospect of harmony amongst colourful, fun-loving, indolent and exotic inhabitants.

This fantasy tourism seems to be largely unimpeded by tricky issues of authenticity. *Treasure Island* continues to be used optimistically by promoters for its association with paradise, despite Stephenson's dystopia being set on a none too pleasant island where life was likely to be "mean, brutish and short", disease-ridden and plagued by treachery and cruelty.

If "paradise" seems *passé* then an alternative attraction can be found in being amongst the first to participate in the opening of the Earth's last remaining "new frontiers". In August 2016, the luxury cruise ship *Crystal Serenity* embarked on a thirty-two-day voyage that would take it through the many islands of the North-West Passage. Just in case conditions presented more ice than anticipated, the tourists were accompanied by an icebreaker carrying two helicopters for the convenience of landing parties. The cruise organisers stressed the importance of inculcating an understanding of nature, history and culture, and the need to make a positive impact on the communities visited en route.

Amongst the settlements visited was Pond Inlet, where 1,000 passengers would meet its mainly Inuit population of 1,400. In November 2014, Martin Fletcher of *The Telegraph* described Pond Inlet as "a troubled place whose inhabitants have largely lost their links to the land and live in a sorry limbo, *caught between two starkly conflicting cultures*". (The emphasis is mine.) [31]

With flimsy, bestilted houses in short supply, with exorbitant prices for all consumer goods, and with domestic violence, malnutrition, obesity, diabetes and school drop-out rates high, and with suicide rates – particularly among young people – more than ten times the national average, here was a culture rendered meaningless by the loss of its hunting traditions and everything that went with that.

For the passengers on *Crystal Serenity*, the top-end price was reported to be $120,000, and many very wealthy people were travelling in great luxury. Tourism here follows on the heels of explorers who brought disease, on traders who weakened traditional subsistence and who brought starvation when prices crashed, on missionaries who undermined traditional spiritual beliefs, and on government policies that introduced an alien system of justice, an education system that divided families, and an insistence that nomads live in permanent settlements.

There are multiple ironies here, amongst a culture whose contacts with "the outside world" have so often had tragic consequences and where impacts have been more often "fatal" than "positive". Now what has been termed "extinction tourism" will enable the very wealthy to observe securely the very poor "at home", their luxury cruise made possible by global warming, to which they are contributing by creating a huge carbon footprint. The Inuit, meanwhile, are faced with their own reality, one in which global warming opens them up to the impact of a race for resource exploration and exploitation. History suggests this could indeed be fatal. [32]

The world created for tourism often does not reserve much space for ethical considerations. Thus Kathy Marks describes the atmosphere of reality suspended when an American cruise ship arrives at Pitcairn Island during the trial of seven of the island leaders accused of rape and child abuse: "Let's make believe that everything is rosy in the legendary island of *Bounty* fame. Let's make believe that the Pitcairn Islanders are all fine, upstanding citizens. Let's make believe that half of the mutineers' male heirs, including our lunch host, aren't accused of sex crimes. ... It's not everywhere that crowds of tourists rub shoulders with alleged paedophiles, visit them in their homes, buy their souvenirs, pose for photographs with them, and generally treat them like nobility." [33]

Chapter Five

Political Dependence
and Turbulence

If you are hoping to find peaceful living under caring governance on a small piece of land in the middle of the ocean, you may well be disappointed. Indeed, it seems you are more likely to find a self-appointed demagogue than an egalitarian utopia. And so, in places as diverse as St Kilda, the English Channel Islands, Tristan da Cunha, Easter Island, Pitcairn Island, Clipperton Atoll, the Galapagos Islands, and Papua New Guinea, men and women have come ashore and proclaimed their right to govern. Some were just opportunistic conmen running from another place, some were perhaps genuine in searching for fertile ground where religious and social ideas could be brought to life. But they all shared the fact that their entitlement to rule was based on nothing but bluff.

Unusual Rulers

Bogus proclamations of a right to rule have been visited upon a diversity of island settings. On St Kilda, "Roderick the Imposter" established a theocracy at the end of the seventeenth century, taking advantage of the remoteness and established religious inclinations of the population. In the English Channel Islands, a fisherman gave himself the title of "King of the Ecréhous" in 1848, and acted accordingly. His position was perhaps tacitly accepted by Queen Victoria when she visited Jersey in 1857. [1]

Far away, and five years before Tristan da Cunha was annexed by Britain in 1816, the self-styled "Emperor of Tristan da Cunha" took it upon himself to establish this most isolated of bogus empires. Jonathan Lambert was an American adventurer-cum-pirate who landed in 1811 with five crew members and, so it is reputed, a pile of treasure. On the remotest inhabited archipelago in the world, he planned to build a trading station and renamed it The Refreshment Islands; but he soon realised the venture was doomed, and Lambert dispatched a begging letter to the Crown asking for rescue. There was still no official answer three years later when a British ship arrived to find the Italian Tomasso Corri was the only living soul on Tristan. He was suspected of murdering his business partners over lost treasure, but a more probable if less exotic explanation is that Lambert was killed in a boating accident in 1812.

Sixty years later, on Easter Island, the French sea captain Jean-Baptiste Onésime Dutrou-Bornier proclaimed his autocratic rule, which included transporting many indigenous Rapanui from the island. Meanwhile, about the same time, the self-proclaimed "Charles I, Emperor of Oceania", was setting out his domain in southern New Ireland, Papua New Guinea, in the 1880s. [2]

In 1832, Joshua Hill landed on Pitcairn Island and informed the islanders, who had petitioned the London Missionary Society to send a qualified teacher and preacher to lead the community, that he had been sent by the British government to take charge of affairs. He had not been sent by the government, nor was he from the London Missionary Society. Instead, Hill had travelled South America and Pacific islands as an adventurer and confidence man. Well educated and claiming links with English aristocracy, he was well versed in duping vulnerable remote communities into falling under his spell. [3]

Most of these "empires" were relatively short-lived, if sometimes catastrophic for the communities where day-to-day existence was subject to an ill-defined and perhaps summary rule of law. That outsiders could hold sway in this manner was a factor of remoteness that ensured visits from the outside world were rare. It depended too on the negligence of colonising powers, which were often only too happy to let their distant charges get on the best they could without interference or the financial cost more responsible governance might involve. And, feeling neglected and perhaps already at the mercy of poor or corrupt local leadership, islanders could easily fall prey to a strong man, the likes of whom they had never seen before.

However, some of the unconventional systems of government have a much longer history. Anachronistic *seigneurial* rule on Sark, in the English Channel Islands, dates back to 1563. With a population of around 600, it was a crown dependency with its own judiciary and legal system. Until very recently only twelve of the sixty-two local representatives were elected, the rest appointed through the *seigneur*, whose position was passed on through inheritance. So governance had a certain feudal quality to it. [4]

The power of money has enabled democracy to be translated into autocratic governance in some instances. In recent times in the Caribbean, the large and wealthy Bird family dominated Antiguan politics from 1981 when Vere Bird was elected Premier, until 2004 when his son Lester Bird was voted out from national leadership. Run like a family business, according to some critics, they survived accusations of corruption and the abuse of authority, particularly in relation to investment in tourism infrastructure, and of harbouring underworld figures and smuggling arms to Colombian drug cartels.

Elsewhere, paternalism has managed to keep democracy at bay. In 1827, John Clunies-Ross settled in the Cocos

(Keeling) Islands, a collection of twenty-six coral cays in the Indian Ocean, just fourteen square kilometres in size and some 2,750 kilometres north-west of Perth. Subsequently, the islands were granted to the Clunies-Ross family "in perpetuity" by Queen Victoria in 1886, and this first generation established copra plantations on the purportedly unpopulated island group, shipping in indentured labourers from the East Indies. Workers were paid in plastic money redeemable at the company store, and the family provided education. Indeed, the fourth generation Clunies-Ross taught local children, performed medical and dental work, and presided over his own court. The family insisted on naming every child born on Cocos (Keeling). Those who reacted against these arrangements and chose to leave the islands were never allowed to return.

Of this system, the fifth generation John Clunies-Ross Jnr reckoned that, through paternalism, the community secured many of the aims of a secure social democracy: full employment with paid holidays, full healthcare and a pension system. Later, he claimed, as democracy was instituted, much of this social welfare was swept away. [5] The impetus behind this change was an Australian government-sponsored United Nations mission to the atoll that denounced relations between the family and their workers as anachronistic and feudal. Not surprisingly, this was in line with Australia's intention to assume sovereignty of the islands, which they did in 1955; whereupon the fourth generation Clunies-Ross avoided compulsory purchase by selling the island group for AU$6.25 million. In a United Nations-sponsored vote in 1984, the islanders opted for full integration with mainland Australia, and so went from living under paternalistic absolute rule, with one local family controlling almost every aspect of their lives, to living in an external territory of a nation thousands of kilometres away, subject to Western Australia state laws and represented in Canberra by a Northern Territory politician. It is perhaps not surprising that, while the islanders are no longer patronised absolutely, they feel they are now being neglected. The 500 or so "Cocos Malays", descendants of the indentured labourers, live on Home Island and the majority are dependent on welfare payments. John Clunies-Ross no longer lives in the family ancestral mansion, but the family keeps a bungalow nearby. West Island is home to about 100 Europeans, most of them contracted by the government to provide services for Home Island. The copra industry failed not long after Australia took over, and it is estimated that it costs about AU$40,000 per person annually to provide services.

But it is unlikely Australia will quibble at this because, more important than all that has gone before, Cocos (Keeling) is considered strategically important. It is located at the edge of Asia, the new centre of global economic growth and the focus of geopolitical manoeuvring by all the major powers. And it has an airstrip that can accommodate large military planes, a key factor in current negotiations between Australia and the United States.

Independent Islands or Dependent Dominions?

Political independence would appear to confer endowments that are disproportionate to an island's size and resource base. States may choose to raise revenue through a variety of sources, to operate free ports, to establish favourable environments for financial institutions and investment, to be the recipient of international grants, and to generally express directly their aspirations within the international community. In addition, where there are resources with high international market value, there is the extra potential for direct capital accumulation which independence endows.

That independent island nations are responsible for their own economic development is an issue that has been the inspiration for many nationalist and independence movements. For example, the Kanak Socialist National Liberation Front has been active in seeking independence in nickel-rich New Caledonia, and in copper-rich Bougainville the Revolutionary Army has been fighting for separatism against Papua New Guinea. Success should enable a newly formed state to prosper directly from its resources but, as we have seen in relation to the economic fragility of many islands, they often need overseas investment to exploit their natural resources and suffer a loss of control and a reduced financial return accordingly. Even then, as we have witnessed in the case of Nauru, long-term prosperity based largely on a single resource is difficult to achieve without diligent strategic management.

It is apparent, too, that islanders are often just as keen to shake off the influence of neighbouring island communities with which they are politically associated as they are to free themselves from their colonial heritage. Entrenched inter-island conflicts inevitably favour political fragmentation rather than coalition, and indeed it has been well observed that islanders are "never happier with their insularity than when asserting that they are completely different from their neighbours, particularly in regard to language, customs and laws, currency, system of government and all other symbols which demonstrate the existence of a small self-contained universe. Consequently, small islands tend to band together only under the pressure of external forces." [6]

It is a consequence of islanders wishing to hold on to their sense of uniqueness that secession movements have often followed upon archipelagos gaining nation status. The Gilbert and Ellice Islands fragmented to Kiribati and Tuvalu, and the American Trust Territories separated into Micronesian nations. Mayotte seceded from Comoros upon independence and became France's 101st *département*, and in 1997 the islands of Anjouan and Moheli declared themselves independent of Comoros in an attempt to restore French rule. This overture was rejected by France, leading to bloody conflicts that culminated in a French-trained former *gendarme* seizing power in Anjouan. Despite numerous reports of human rights abuse, he held sway there until 2008, when African Union and Comoran forces seized it back.

There have been tensions too between Trinidad and Tobago, and between Curaçao and the northern Nether-lands Antilles. Elsewhere in the West Indies, St Kitts and Nevis were established as part of a UK federation along with Anguilla in 1883, until Anguilla declared itself *rede-pendent* on the UK in 1967. St Kitts and Nevis were declared fully independent in 1983, and thereupon Nevis (popula-tion about 9,000) began to nurture feelings of resentment over the big brother attitude of St Kitts (population about 32,000). This resulted in Nevis initiating the constitutional process of secession, which was narrowly defeated in a referendum in 1998, since when Nevis politicians seem to have worked more towards constitutional reform rather than separation from St Kitts.

The ardour with which islanders once expressed their uniqueness and desire for independence seems to have been cooled by the chilly winds of economic reality. Indeed, it is more than twenty-five years since the United Nations Special Committee on Decolonisation confirmed that none of the remaining island colonies view independence as a desirable option. Most desire a larger degree of self-government and active involvement in their own affairs, but recognise that the *realpolitik* lies in substantial socio-economic benefits derived from political affiliation. These include free trade and export preferences, lucrative grants and social welfare assistance from the parent country, special tax concessions on borrowing, assistance with infrastructure and communications projects and with health and education systems, natural disaster relief, external defence, and even protection from internal disturbance.

In the aftermath of energetic post-World War II activity to gain independence – in the Caribbean, for example – there remain sixteen dependent island democracies that, almost without exception, have indicated through the ballot box that they wish to remain so. These dependencies are significantly small, averaging less than ten per cent of land area and population compared with their autonomous neighbours. Despite this, the dependent territories register higher GDP and electricity production, higher GDP growth, higher motor vehicle ownership and telephone subscription per capita, lower unemployment, infant mortality and staff-student ratios in education, and higher life expectancy, higher results in standard education tests, and larger numbers of doctors and hospital beds per capita. [7]

However, this relative socio-economic advantage rests on an insecure footing. Their small size and relatively high populations make them more vulnerable to natural disasters like hurricanes. With fewer alternatives, their economies tend to be dominated by the tourism industry, which, for reasons already discussed, is often unstable and cyclical in nature. They are also more involved in off-shore financial activities, which are dependent on volatile international market forces and decisions made anonymously and thousands of miles away.

They are vulnerable also to the increasingly competitive forces of economic globalisation. Major economic powers are rather disinclined to be generous with ex-colonial outposts unless there are specific and significant political and economic incentives. This is the case with Britain's involvement in the Falkland Islands, discussed in detail later in this chapter. Rather, they exhibit a keen self-interest in profiting from a free market uninhibited by tariffs and where preferential trade agreements are skewed towards themselves. During times of fiscal and economic woes, when it sometimes feels that the future of the capitalist socio-economic enterprise is in doubt, nations like Britain and France will favour popular and politically acceptable mainland priorities at the expense of taking care of island remnants of colonial history. The fact that these outposts are for the most part many thousands of kilometres away and rarely in the public view unless there is a natural catastrophe or a scandal, this could allow them to drift further off the political agenda, far enough to be out of sight so out of mind.

They are dependent too on the level of competence and financial commitment exercised by a distant sovereign nation perceived as anonymous and ill-informed. The quality of responsibility exercised by the mother country is always likely to be uncertain, and these children of latter-day empire can be sorely neglected, as the Foreign Office in London deals in *realpolitik* not in altruism, and it is the same in Paris as it is in Washington.

So hard-nosed British foreign policy determined that the people of the Chagos Islands could be cleared, transported, deported – "swept clean" in the parlance of the day – to make way for a US base; while the Falkland Islands are taken care of for their varied coastal wildlife, rich fisheries, significant potential for offshore oil, and their ability to rouse the hearts of British empire loyalists in populist sabre-rattling against Argentine sovereignty claims.

The Falklands Factor

Events seem to happen quickly in the turbulent political and economic history of some islands. Just as tiny Nauru in the Pacific went from having per capita wealth greater than that of the US and Saudi Arabia in 1993, and was bankrupt by 2004, so events have gone the other way in the Falklands. In 1985, the journalist Simon Winchester described the Falklands as having very little local industry, and half-a-million sheep but no wool processing enterprise. Nearly half the grazings were run by a company owned by Coalite with land divided up into farms run on a near-feudal system by absentee landlords. The fishing industry was more a matter of prospect than reality, and tourist-related projects were based on a landscape with no trees, with bog and bare rock stretching into a distance so often obscured by windswept rain and salt spray. [8]

Since then, the "Falklands Factor" has been brought into play, whereby an overseas crisis diverts public attention from more pressing domestic matters – arguably a major reason behind the UK going to war with Argentina in 1982. The sovereignty battle was won, for the moment, but tied to this, and increasingly important in the aftermath, was a belief that pouring resources into the islands could have significant economic consequences. By 2010, GDP had risen from £5 million to £105 million, population had risen by sixty-five per cent, and in a 370-kilometre Economic Exclusion Zone fishing was flourishing, especially through the sale of licences to Korea, Taiwan, Russia and Spain for squid fishing – a transformation described as "sheepocracy to squidocracy". There has been something of a social, political, economic and cultural revolution, with new communications, schools, hospitals, dockyards, the prospect of super-casinos, a new £350 million air base, and full employment. Tourism accounts for more than 100,000 visitors a year, with the cruise ship market flourishing from October to April when visitors anticipate seeing many varieties of sea lion, whale, seal and penguin and a wide range of avian wildlife. The war has even rejuvenated an interest in military history.

Most importantly, recent explorations have indicated that oil reserves in the North Falklands Basin are worth in excess of £60 billion. With all this going on, it is not surprising that there are claims of economic self-sufficiency in the future – something dear to the heart of British politicians. Given that the cost of the military garrison is in excess of £60 million a year, and that UK grants totalled £46 million in 2011, it is difficult to believe that self-sufficiency can be achieved other than by having high, perhaps inflated, expectations that future oil and gas finds can be profitably exploited.

The transformation appears remarkable, and is not something that sits easily with Argentina. Diplomatic ties were severed between Britain and Argentina after the 1982 conflict, and restored in 1990 based on an "umbrella formula" – a pragmatic quasi-arrangement to agree to disagree about sovereignty and to get on with other business. This enabled, for example, Britain to lift its trade embargo and the ban on Argentine private aircraft and vessels visiting the islands; there was also a joint plan to co-operate on oil exploitation in 1995, and an agreement to share data concerning sustainable fishing. But these protocols have been repudiated and then reconsidered by successive governments in Argentina, and the issue of sovereignty has become even more intractable in recent years, with both sides ever more polarised in the righteousness of their cause. And as one observer explains "once rumours of oil seep out, they poison negotiations just as surely as slicks blacken the sea. ...When the islands were unimportant, their fate would have been easy to settle. The magnitude of the problem has grown with the magnitude of the stakes." [9]

Meantime, Argentina has continued to complain about the injustices of Britain's original seizure of the islands in 1833, and to assert – probably correctly – that Britain is making an erroneous claim at the United Nations in arguing that the islanders are exercising "self-determination" in choosing to be colonised by Britain rather than Argentina. Indeed, the UN will only support self-determination when a country seeks decolonisation and independence from a colonising power, which is not the case with the Falklands.

With an increasingly valuable fishing industry, with potentially rich oil resources, with very substantial grants and infrastructure support from Britain and with the 1,400 inhabitants being protected by an almost equal number of service men and women, it is hardly surprising that news headlines have focused on Falklanders being "the luckiest working-class people in the world". [10] Inhabitants of Britain's other overseas territories may have good reason to consider themselves less lucky.

Colonial Remnants

Beyond a scattering of about 200 island "outposts" that London administers, there is little else left of the British Empire – the peninsula of Gibraltar being the largest non-island exception. Simon Winchester visited almost every British island possession – the British Indian Ocean Territories, Tristan da Cunha, Ascension, St Helena, Hong Kong (still a colony in 1985), Bermuda, the remaining British West Indies and the Falklands – and in most cases spoke with the Queen's representative. His critical conclusion was that the islands were "trapped in history, condemned to an eternity of begrudged expenditure, parsimonious direction, second-rate thought and government, listlessness and ill fortune". [11] (It should be noted that this view of London-based administration is not restricted to Britain's far-flung remnants of empire. In pre-devolution days, politicians in Scotland and Wales often accused Whitehall of benign indifference.)

St Helena, for example, is about 120 square kilometres,

3,100 kilometres from South America and 1,850 kilometres from Africa. It is a five-day boat journey from the nearest international airport in Cape Town. Once, thousands of ships travelling between Europe and Asia and South Africa called to take on fresh water and dump scurvy victims. Some of the islanders are descended from slaves, and another group is descended from Londoners made homeless in the Great Fire of 1666. The travel writer Harry Ritchie visited in 1997, when the island barely received 300 visitors a year, and described it as "a depressing, humiliating experience. The Saints have always been devotedly patriotic and very warm hosts. There are pictures of the Royal Family in every sitting room. But it is obvious Britain couldn't care less about the colony." [12] With the traditional hemp crop redundant since the Royal Mail switched to synthetic string, with virtually no private industry, no regional market and no outside investment, at the time of Ritchie's visit island imports were valued at £4.7 million, and exports – mainly postage stamps, frozen tuna, and very expensive coffee and honey – were worth just £145,000. Unemployment was running at eighteen per cent, and the maximum government benefit was £40 a week.

Many Saints work on Ascension – where they constitute about three-quarters of the population – and the Falklands. Having lost their British citizenship and right of abode as a result of the Nationality Act in 1981 (passed to satisfy British politicians that millions of panicky Hong Kong people could not "invade" during the process of the colony being handed over to China), and having regained it in 2002 when the act was revoked, critics have argued that St Helenians have been rewarded for their loyalty by being ignored and have little choice but to become a kind of mobile workforce for the South Atlantic. [13]

During elections in 2002, the governor was attacked, and two of the executive council of five resigned amidst claims that the governor was acting like a dictator. Tensions have relaxed with the introduction of a new constitution, but this has done nothing to stem St Helena's steady population decline from 5,200 in 1998 to 4,250 in 2008.

Much was at stake in the British government's decision to build an airport, completed at a cost to Britain of £285.5 million. It was anticipated that, in the long term, the airport's link with South Africa would reduce the £26 million annual aid bill by providing employment and by increasing tourism – an activity almost entirely built around the site of Napoleon's imprisonment – to 30,000 visitors a year, and so render the island self-sufficient. It was always a hugely optimistic plan and has been rendered even more fanciful by the fact that, at the time of writing, no commercial flights are able to use the airport. The cause is turbulent, squally and difficult to predict gusty winds bouncing around two nearby mountains. This was something noted by Darwin in the early nineteenth century and was identified in an airport feasibility survey, but it was seemingly ignored by Britain's Department for International Development. The only decision that seems to have been made so far is that a solution does not lie in blowing the tops off the offending mountains.

The remnants of French empire are almost all islands, too. These include French Polynesia (including Tahiti, Mururoa and Fangataufa), New Caledonia in the western Pacific, Wallis and Futuna in the central Pacific, Mayotte off East Africa, Guadeloupe and Martinique in the West

Indies, and Réunion in the Indian Ocean. French Guiana, on the north-eastern coast of South America, is a significant non-island exception, but even this *département* includes the Islands of Salut.

Despite having close ties with France, with members being elected to the French parliament, critics have argued that the South Pacific – with its numerous significant independence movements – has suffered from the imposition of a consumer society, with money being focused mainly on urban areas, leading to outlying islands becoming depopulated. An increasing number of urban poor live below the poverty line, and there are inner-city tensions and racial conflicts between Maori, Chinese and French populations in Papeete, the capital of French Polynesia, which culminated in riots in 1987 when one third of the city was burnt to the ground. Sporadic violence has continued since.

The continued French presence on an estimated 120 islands and atolls was justified by the Cold War and its demands for nuclear weapons testing. With a moratorium on testing, most of the French military presence could be redundant, and France might do well to act sympathetically towards independence movements in seeking to offload its Polynesian dependencies. This prospect would place many islanders in a dilemma, and they may decide – like people on many potential island states in the Caribbean – that dependence is a better option.

Revolutionary Rervour
Where powerful independence movements exist, for example in copper-rich Bougainville, which is part of Papua New Guinea, there is usually a key economic advantage

at stake. However, internal civil disturbance in the form of coups has causes not always easy to define. For some island nations, coups seem to be just part of political life. This has been true for the Comoros as it has been for Fiji. For others, like the Maldives, a long-established autocratic regime has used what has been described as a police coup to unseat the short-lived government of the archipelago's first democracy.[14]

The Comoros gained independence from France in 1961, since when the island nation has been subject to some twenty attempted coups. Of these, South African-based mercenaries under Bob Denard were involved in several, the one in 1978 reportedly receiving discreet official support from the French secret service in overthrowing a Marxist leader. In the aftermath, 900 elite French troops were garrisoned in the islands under a defence agreement.

There were times when Denard seems to have run affairs as a personal fiefdom: he took Comoran citizenship, converted to Islam, married a local person, and was the nation's "security chief" from 1978 to 1990, employing about thirty mercenaries to run the presidential guard – highly visible and reinforcing an image of white superiority. The fourth coup in which Denard was involved was repelled by a French expeditionary force, and signalled his demise as a subversive political force, and he spent the next twelve years defending court cases by claiming that he had implicit or explicit support from French security forces for his activities.

In 2008, the new president of the Maldives, Mohamed Nasheed, quickly became well-known as a climate change ambassador for small island nations, appealing to industrialised nations at climate change summits to take responsibility for their role in the potential destruction of island nations. But prior to winning the Maldives's first democratic election in 2008, he had been imprisoned by an administration whose thirty-year rule was marked by corruption and human rights abuses. In February 2012, he was forced from office and by the end of the year, the Maldives image as a tourist paradise was being challenged by human rights investigators, who accused the police of serious repeated civil rights abuses. The situation could have been embarrassing for the British government, which was quick to recognise democratically elected Nasheed (described by David Cameron as his "new friend") and equally quick to recognise the government that forced him out in an apparent coup. Britain has now addressed any "misunderstanding" by granting the ex-president political refugee status. [15]

Chapter Six

No Value But That of Location

Remote, small, sparsely inhabited islands run the risk of being perceived from afar as valuable only for their location. Accordingly, their significance is assessed in a singularly objective manner in the pursuit of specific, externally directed strategic endeavours. These are typically concerned with trade, communication and the siting of military bases, and include small islands used for sailing ship reprovisioning, steam ship bunkering, telegraphic cable stations, long-haul aircraft refuelling, satellite tracking stations, and weapons testing. Indigenous people would either be non-existent, evacuated or ignored. Changes in geopolitics, strategic need and technology determine whether islands are depopulated, evacuated or re-populated as necessary; and so the utility of small, remote islands demonstrates remarkable versatility to changing circumstances.

In these days of remote-control planetary weaponry, one would expect islands to have lost the relevance they once enjoyed, when sailing ships overran the oceans of the world. But this is far from the case, not least because they have the dubious advantage of being isolated from international attention. They can be relatively easily transformed beyond recognition and, in the case of nuclear weapons testing, rendered toxic for thousands of years. And at times of military conflict they can take on a significance that appears entirely disproportionate to their size and location – the Falkland Islands conflict being a case in point. These islands would be perceived as having no value but that of location.

Perhaps the ultimate expression of small islands as location units is the construction of artificial islands suitable for nuclear electricity-generating plants. This was recognised as an option in the early 1980s in a European Commission report. Such islands would have plentiful supplies of cooling water, and – so it was argued – could be sited within an economic distance of electricity load centres, and would offer a degree of isolation because of the surrounding cordon of water – a factor recognised in the choice of island locations for other anti-social industries, like gas and oil terminals and chemical plants. It was envisaged that Britain provided a good choice of locations within twenty kilometres of the coast, particularly off the south-east coast, the Bristol Channel, Wales and Morecambe Bay.

Indeed, there is a precedent for constructing islands around Britain. Seven clusters of Maunsell sea forts – named after their designer and looking like huge steel boxes standing on hollow steel and concrete legs – were sited between Harwich and Margate in the Thames Estuary during World War II. They were built to defend against German planes on their way to London, and to deter mine-laying by enemy submarines. Most importantly as it turned out – because they came into service after the worst of the Blitz – they extended radar coverage to the coast of Europe. For the 3,000 men who served on the forts, this would have been a dreadful posting, with ear-splitting noise when firing the guns, and for the rest of time waiting in a claustrophobic steel box, hoping that a drifting mine, a submarine or a German plane would not strike at the fort's arsenal. It is unsurprising then that suicide and desertion was more destructive than enemy action. [1]

The argument that islands are suitable sites for handling harmful materials of the sort policy-makers prefer to distance the public from is based on the recognition that enisling a problem – locating it "out there", not "in here"

– means that it is conveniently off-shore and thus out of mind. There are limitations in this rationale, but at least it goes some way to explaining why offshore islands will be used for handling toxic cargoes like liquid petroleum gas, and the more hazardous liquid natural gas and nuclear fuel residues. The Canary Islands and the Cayman Islands are involved in these activities; and Taiwan ships its radioactive waste to Orchid Island, populated by the rural, ethnically distinct Yami people, who now share their island not only with its beautiful eponymous flora and flying fish, but 100,000 barrels of radioactive waste.

Palau, in the western Pacific, long resisted overtures from the United States to militarise, which would involve hazardous chemical handling and weapons storage on the island. The United States has been criticised for the manner in which it overcame this opposition, but in fairness to them, their policy-makers are not averse to looking closer to home for these grim enterprises. In 1997, it was reported that the US Army Corps of Engineers was searching for a site to store industrial waste, one option being to build a 1,000-hectare contaminated waste island between Staten Island and the New Jersey shore. It was envisioned that this toxic mud island would be bigger than Coney Island. The option was not taken up.

Pawns in a Geopolitical Game

In the relationship between powerful states and their dependent possessions, islands have been annexed and traded at will, treated as disposable and exchangeable units as nations have striven for regional power and influence. Mauritius experienced eighteen Dutch, twenty-one French and thirty-one British governors before gaining independence. The Falklands were held by France, Spain (twice), Britain (twice) and Argentina before being "secured" by Britain in 1833. Fernando de Noronha, off the coast of north-east Brazil, first appeared on a Portuguese map in 1502, since when it has been held by Britain, France (twice), the Netherlands (twice) and finally by Portugal, who held it by building ten forts. Its pre-eminent use was as a penal colony, it was a military base in World War II, and has been used by the United States as a missile-tracking station. [2]

In the Caribbean, there was something of a "fashion for islands", during the first half of the nineteenth century, Britain expanded trade through bases in Barbados, Nevis, Montserrat and Antigua. They began too the transformation of Bermuda, hitherto considered to be God's perfect garden.

By 1885, the Anglo-Chilean diplomat Vicuña Mackenna could reflect how this "fashion" had been extended globally. In the context of the political future and strategic significance of Easter Island he observed that "Great Britain, like the child-eating giants of the stories, is an insatiable island-eating nation ... which, stimulating its appetite with an archipelago such as the Bermudas or Falkland Islands, follows by digesting the New Hebrides or some delicious morsel like Cyprus." [3] This hunger is probably the product of a kind of psychostrategic anxiety shared by colonising powers – the recognition that if a state does not claim an island then another state may intercede and claim it for themselves.

On the other side of the world, island-eating is evident too in the Kurile Islands between Japanese Hokkaido in the south and Russian Kamchatka in the north. These were first partitioned between Japan and Russia, then in 1875 the whole chain was traded to Japan in exchange for all of Sakhalin Island, which subsequently changed hands again during the war of 1905 and the Russian Civil War. The Kuriles remained Japanese until 1945, when Stalin made a quick attack and took southern Sakhalin and the Kuriles even as Japan was on the point of surrender. The Russian presence, just two kilometres from Hokkaido, continues to disrupt trade negotiations between the two nations, but despite being keen now to foster good relationships with Japan, Russia seems equally determined to protect the Kuriles as the gateway for their atomic submarines passing between the huge naval base at Vladivostok and the North Pacific.

The Kamchatka Peninsula is highly militarised and is home to one of Russia's principal rocket test firing sites and naval bases. From here submarines set out on patrols of the Atlantic and the Pacific, monitored by American military bases on Aleutian islands in the Bering Sea, such as Shemya, Amchitka, Adak, Attu and Kiska. [4]

The nationalist sentiment that has driven groups in Japan to demand from Russia the return of the southern Kuriles is reflected too in China where, with rare leadership uncertainties, rumours of political corruption, and with rapid economic growth no longer taken for granted, one way of reaffirming the legitimacy of communist rule is by appealing to the popular nationalism that territorial threats evoke. This apparent revanchism also serves to warn the United States to stay out of disputes within China's increasing sphere of influence.

In the East China Sea, the Diaoyutai chain (Senkaku to the Japanese) is made up of eight small deserted islands about 150 kilometres north-east of Taiwan. They were first controlled by Japan in 1895, and it was only in 1968, when it was discovered that there could be oil reserves in the area, that China raised sovereignty issues over Japan's obligation to hand back territory after World War II. The islands are also claimed by Taiwan. In 1990, they were the site of seemingly bizarre nationalistic flag-waving, slogan-painting and torch-planting ceremonies.

However much this may appear to be merely the rhetoric of nationalism, there is a much more serious game being played out here. With the discovery of rich oil and gas deposits, political tensions have increased to new levels, just as political alignments have been re-evaluated. Indeed, by September 2012, and after Japan had purchased some of the islands from private owners, the British newspaper *The Independent* was describing the Senkaku/Diaoyutai Islands as barren rocks dividing the ambitions of superpowers. [5]

The dispute escalated when protestors attacked the Japanese embassy in Beijing after some forty Taiwanese fishing boats and twelve patrol boats exchanged water cannon fire with Japanese coastguard ships. And the economic impact of conflict with its largest trading partner has been significantly detrimental for Japan. Amongst several internationally known companies – including Canon, Panasonic, Honda, Toyota and Nissan – profits were affected considerably, with millions of shareholders made anxious and made aware of these few uninhabited barren rocks.

For countries bordering the East China Sea, the South China Sea and the Yellow Sea, geopolitics is convoluted, confused and increasingly recognised as a serious threat to world security. As we have seen, China and Japan have a tense relationship in determining island sovereignty, and China's relationship with South Korea has soured with Seoul's decision to deploy a US missile system on an island aimed at North Korea.

In July 2013, it was reported that Japanese and US troops were in combined exercises practising retaking a remote island airport from enemy forces, whilst China and Russia were involved in their largest ever joint naval drills. In December 2016, Vietnam took part in naval exercises with the Philippines, who a month later engaged in joint operations with Russia, too. The Filipino navy is virtually non-existent, and the country has a defence treaty with the United States. This is the culmination of a 70-year alliance, but President Rodrigo Duterte has expressed openly his wish to break the link and realign with China – something the Chinese government commends even as it encroaches upon and colonises Filipino island territory – whilst reaffirming his commitment to treaty obligations with the United States.

China has declared an Air Defence Identification Zone in the East China Sea which includes the Diaoyutai/Senkaku Islands and within which all foreign planes face the prospect of Chinese fighter planes taking "defensive emergency measures" as they see fit. Japanese, South Korean and US aircraft have been making a show of ignoring the zone, whilst some international commercial operators have avoided the area. As the US State Department has noted, this heightened regional tension involving five nations, each with a high military capability, increases the risk of miscalculation, confusion and accidents.

Islands in these seas have been squabbled over for

centuries on account of abundant fishing grounds, but high expectations of potential oil and gas exploration have upped the stakes to the extent that some of the smallest islands on Earth are at the centre of the world's biggest territorial disputes. China, Vietnam and the Philippines have been involved in angry standoffs around the Paracel Islands and the Spratly Archipelago since 1974. The discord has led to violent provocations and military skirmishes resulting in the death of nearly 150 military personnel, multiple ship collisions and the sinking of three ships, the claimed sabotage of two oil exploration operations, and the involvement of the United Nations Convention of the Laws of the Sea. In May 2014, China positioned a giant oil rig near the islands, just 250 kilometres from the Vietnamese coastline, provoking street protests and the closure of Chinese-owned factories in Vietnam.

Currently under Chinese administration, Taiwan and Vietnam claim the Paracels, but it is unlikely China will relinquish them as they protect the route of its nuclear submarines stationed on Hainan Island. On the Spratly Islands, consisting of over 100 small islands and reefs, China, Taiwan and Vietnam claim the entirety of the archipelago while Malaysia and the Philippines also claim parts. And Brunei has established a fishing zone based on a southern reef, but has made no formal claim to the territory.

China lays claim to most of the South China Sea. If this claim is based on anything it is the hugely over-stated significance of the strategic control it gained briefly in the early fifteenth century when its admirals were involved in a flurry of activity from South-East Asia to Africa. All other interested maritime parties challenge this claim as illegal.

The Island

A United Nations Convention of the Laws of the Sea (UNCLOS) arbitration ruling in October 2015 declared that a case brought by the Philippines against China deserved determination, and it has since ruled in favour of the Philippines. This leaves China to ponder the prospect of little more than a twelve-nautical-mile territorial sea entitlement around four "rocks". Land deemed to have elevation only at low tide *in their natural state* has no bearing on determining maritime boundaries.

Vietnam, Malaysia, the Philippines and Taiwan have all expanded islands in the Spratlys, but it is China that has engaged in island building on an industrial scale. Using dredgers to pump millions of tonnes of sand onto and around coral reefs, this becomes the foundation for raising huge concrete structures. Fiery Cross Reef, for example, was a few rocks barely visible at high tide, but has been "developed" into a 270-hectare island with a three-kilometre runway that can accommodate military and commercial jets. Its port can accommodate military tankers, and it has two lighthouses, helipads, a plethora of communications structures and extensive support facilities. Although China has made a commitment not to militarise its controversial island constructions, a US think tank has presented evidence of defence platforms – anti-aircraft guns and missile defence systems – on all seven of the islands it occupies in the Spratlys. [6]

At the time of writing, China simply denies the validity of the UNCLOS ruling, and if she continues to behave as if the Spratly Archipelago is a "core interest" – like Taiwan and Tibet, over which she brooks no international opposition – then the present scramble for political alliances will only accelerate.

No Value But That of Location

The South China Sea is one of the most important maritime routes in the world. Ship-borne trade is reckoned to be worth £3.2 trillion, and so it is unlikely to become a "Chinese lake" without a huge struggle. The United States has not formally signed UNCLOS, but operates unilaterally a Freedom of Navigation programme through which it asserts the principle of navigation in the sea and challenges what it perceives as the excessive national claims to international global ocean and air space. In pursuing these aims in late 2015, the missile-carrying destroyer USS *Lassen* sailed into waters claimed by China – a presence reinforced most effectively by flying two B52 bombers over Chinese island-building operations just ahead of a government summit in Manila to discuss joint interests in the region.

China claims that the United States is using these South China Seas territorial issues as a means of corralling countries who oppose China's claims into joining forces and containing her increasing economic and military growth, and indeed security analysts have recently described a shift in US policy away from the Middle East. It is perhaps not coincidental that oil resources in the South China Sea have been estimated to run as high as $1 trillion.

The United States is concerned that China is redefining international law to suit its interests alone, and in this it is certainly finding allies. Japan and South Korea depend on passage through this area especially for their oil and gas supplies and, as we have seen, Japan has its own conflicts with China over Senkaku/Diaoyutai and so seeks cause with Vietnam and the Philippines in supplying marine defence equipment, just as South Korea provides weapons to Indonesia and the Philippines.

The Island

India has historical enmity with China, and is concerned about the island state alliances China is forging across the Indian Ocean – the so-called "string of pearls" – and is seeking security ties with Vietnam, Indonesia, Japan and Australia. This maritime area is a crucial transit zone for trade between eastern Asia and the Middle East, and China has negotiated, financed, constructed and made foreign aid contributions in order to establish port facilities in Myanmar (Burma), Pakistan, Sri Lanka and Bangladesh. And in 2011, it established a naval base in the Seychelles. This has not been to the liking of India, suspecting as it does that they may be strangled out by China's increasing influence across the Bay of Bengal and beyond. This anxiety has reverberations in the unstable political situation in the Maldives (see Chapter Five). India has nervously watched China courting the Maldives with a defence agreement at the end of the 1990s, and with offers of cultural co-operation, favourable trade agreements, and with infrastructure and housing projects by state-aided Chinese contractors.

In the Yellow Sea, disputes between North and South Korea have rumbled on since the end of the Korean War. In 2010, the conflict escalated with the disappearance of a South Korean warship, which was widely blamed on the North Korean navy. Subsequently, the North shelled the disputed Yeonpyeong Island, leading to four fatalities and an increased South Korean military presence. An extension of bases in this area would enable the United States to reinforce its influence in this perennial conflict, which in 1948 saw South Korea, apparently with the tacit approval of the United States, instigate a "Red Cleansing Campaign" on Jeju Island, 100 kilometres south of the South Korean mainland.

Estimates suggest that as many as 30,000 people, ten per cent of the population, were murdered in a brutal precursor to the Korean War. Jeju now exists as an autonomous province. Its coastline is considered to have a unique underwater habitat, and agriculture, tourism and fishing have helped sustain a traditional way of life. The islanders have always opposed the establishment of a US base, but despite this South Korea has been persuaded to help construct one of the largest bases in the world, a provocative nuclear facility less than 500 kilometres from the Chinese mainland.

Surreal Disputes

In the Arctic, Canada has adopted a more bullish attitude to its sovereignty claims in recent years. As early as 1953, Inuit families were moved from northern Quebec settlements to Grise Fiord on Ellesmere Island and to Resolute on Cornwallis Island. Later the Inuit sought compensation for human rights abuses suffered during these resettlements, but the government's priority was to "maintain a presence" on these islands as a way of flagging up a sovereignty interest.

Confrontation between disputing nations in the Arctic – Canada, the United States, Norway, Russia and Denmark (through its control of Greenland) – is spurred by global warming making northern waterways less encumbered by ice in the summer. The North-West Passage first became fully clear of ice in Summer 2007, and already luxury liners – accompanied by vessels able to break through residual ice – are offering hugely expensive "expeditions" from Alaska to New York. Most importantly, as the North-West Passage becomes less a fable than a reality, so the exploration for oil and gas resources becomes credible, too.

Canada, in particular, has public support for pursuing a firm line over Arctic sovereignty; but confrontation between some of the competing nations – which often have political relationships that are otherwise generally sympathetic – can take on a theatrical air. For example, in 1973, Canada and Denmark failed to agree on the boundary between Ellesmere Island and Greenland, a fraction of which is occupied by Hans Island, a tiny rock islet of 1.3 square kilometres. In 1984, a Danish minister landed on the island and planted the Danish flag, and in 1990 the Canadians responded, rather tardily, by replacing the flag with their own and following this up with a visit from their defence minister. In both cases, the contending nations saw fit to leave a bottle of spirits as well as their flag. Negotiations continued at the United Nations, and in April 2012 it was reported that both countries were considering a plan to divide Hans Island down the middle. [7]

This rather surreal quality of small island geopolitics is evident elsewhere. In the early 1990s, Canada was in conflict with the French government of the St Pierre and Miquelon Islands, which lie about twenty-five kilometres off the coast of Newfoundland; and it was in conflict with the 6,500 island inhabitants too, some 2,500 of whom were involved in traditional cod fishing. Canada has wanted to improve conservation measures around its coast and to allow France a twelve-nautical-mile fishing limit only. France is seeking a 200-nautical-mile fishing zone (its allowable Exclusive Economic Zone) around the islands, giving them control over fishing quotas and ownership of potentially huge oil and gas resources. Meanwhile it was reported that the islanders, recognising that the day-to-day *realpolitik* was fish, ignored Canadian-imposed fishing quotas, and in angry opposition to the French government seeking to send factory ships to the area, they forced a military aircraft and its *gendarmes* to return to Paris and set adrift a French naval vessel after pelting the crew with fish-heads!

There was a touch of theatrical drama too when Moroccan frontier guards landed on Perejil Island, less than a kilometre off the Moroccan coast, with baggage consisting of not much more than a radio, some tents and some flags. Moroccan officials claimed, in the furore that followed, that they were setting up an observation post to counter illegal immigrants and potential terrorism from Africa to Europe. This may have satisfied and even given encouragement to the international community, but not to Spain who, even whilst admitting the island had no strategic significance and there were vague uncertainties over its sovereign status, saw fit to describe events as a serious situation and a violation of Spanish territory.

What is serious, but still with a strong comic element, is the quarrel between Greece and Turkey over the sovereignty of close offshore islands in the south-east Aegean. There are profound historical antagonisms between the two nations, with the focus in recent years being divided Cyprus. Even as recently as 1994, a former US Assistant Secretary of State could describe Cyprus as "the hottest spot in the world", reflecting a Greek Cypriot plan to buy forty Russian missiles capable of striking deep into Turkey's airspace and fears that Ankara could launch a pre-emptive strike. The international community is united in working to ensure peaceful relations between the two nations, so when a dispute over a few tiny rocky islands becomes inflamed, the issue takes on a significance that at first sight, and like so many island affairs, seems out of proportion in relation to what is in effect just a speck of land.

When, on December 25, 1995, a Turkish cargo ship, carrying concrete, ran aground on Imia (known as Kardak in Turkey), the Greek authorities were informed that the problem need not concern them, since the island was in Turkey's territorial waters. The mayor of nearby Kalymnos was furious, and planted a Greek flag on Imia; whereupon some Turkish journalists flew in and replaced it with the Turkish flag. Then both the Greek and Turkish navies arrived. The mayor, unrepentant despite the rapidly developing international crisis, declared that "this islet, and some other islets, are Greek according to international treaties, and when something belongs to you, even if it is just a rock, you try to protect it, as you do with your own yard. And Imia is for us our own yard."

At one point, a team of Greek and Turkish commandoes each occupied one of Imia's two rocks (during which three Greeks were killed when a helicopter crashed), and the soldiers were only pulled back after a US mediator intervened. Apparently, the conflict dates back to 1947, when a treaty ceded the islands to Greece after they had been occupied by Italy. Turkey asserts that the wording of the treaty has been misinterpreted, that the Greek community of Kalymnos is not very adjacent to Imia but nine kilometres away, whilst the Turkish Bodrum Peninsula is seven kilometres distant. Turkey also asserts that Imia is not an *island* but a *rock*, a difference that the Turks feel is significant. The events stirred one senior British politician to remark, "We saw over the Imia dispute, for example, how a disagreement over a rocky, uninhabited islet, almost led to a war in the eastern Mediterranean." The issue remains unresolved.

Useful Islands

Islands are the focus of resource competition all over the world, and fish is just one of those valuable assets nations consider worth facing up over. Migingo Island is on Lake Victoria on the border between Kenya and Uganda. It is little bigger than half a football pitch, but has a population of about 500, which must make it amongst the most densely populated places in the world. What appears self-evident, and what research confirms, is that Lake Victoria is dying. The water level is estimated to be dropping by about one metre in ten years, the lake is surrounded by thirty million impoverished people, and there is unregulated fishing, industry and agriculture. This, together with unrestricted sewage disposal, has caused eutrophication (excessive rich-ness of nutrients) and related high acidity rates, which combine to favour invasive species of water hyacinth to form massive mats on the water. With fish stocks down by around seventy per cent, the waters around Migingo Island are the last remaining productive area in the lake, contributing enormously to a Nile perch industry worth £60 million to Uganda in exports. Little more than ten years ago, the productivity of the waters was only a rumour, but one that encouraged fishermen to come in increasing numbers from Kenya three hours away, and from Uganda six hours away.

It was not long before easy pickings were seen to be had by criminals among a small vulnerable population, and for a time a brutal and corrupt group held sway. Ugandan and Kenyan military became involved to restore security for the islanders and to assert sovereignty, which culminated in an international standoff. And this is an area less than half a hectare – "a slab of rock ... packed with rusty metal shacks, heaps of rubbish, glassy-eyed fishermen and squads of prostitutes" the fate of which has "caused outrage in east Africa, [and] triggered a ministerial crisis and brought Kenya and Uganda to the brink of a shooting war". [8] Order was restored when, in May 2009, the Ugandan government conceded that Migingo was in Kenya by a distance of just 510 metres, and they withdrew troops and police; but they continued to argue that Kenyan fishermen were illegally fishing Ugandan waters nearby.

Elsewhere, Chile and Argentina are in contention over possession of Lennox, Picton and Nueva, south of the Beagle Channel. Their value lies not as windswept, rather barren islands, but in their usefulness for claiming sovereignty over potential offshore oil and gas resources. As we have seen in our discussion of Exclusive Economic Zones, the possession of these islands could have a disproportionate effect on the possession of potential deep-water resources further south. Consequently, even as far south as Cape Horn Island, and further offshore on the Diego Ramírez Islands as well, small Chilean bases monitor not only the weather, but also Argentine ship movements and communication channels. [9]

Some 800 kilometres south of Cape Horn, on the few square kilometres of ice-free land on King George Island, there are nine nations running so-called scientific bases, despite the fact that it is questionable whether the area holds anything to explain such intense research interest. The significance of King George Island is that it has been the only place one can reach on the Antarctic continent without a steel-reinforced ship.

Antarctica is the driest, coldest and windiest continent. It is not a country – it has no government or indigenous population – and under the Antarctic Treaty, which came into force in 1961, it is administered as a self-regulating scientific preserve based on international co-operation. Signed by fifty states and up for renewal in 2048, military activity is banned and, in theory, science guides all activity.

A closer look reveals there are approximately seventy bases representing some thirty countries. About twenty-five are on islands. Until 2048, territorial claims are in abeyance, with the continent divided up longitudinally between Norway, Australia, France and New Zealand, with Argentina, Britain and Chile sharing overlapping claims on the Antarctic Peninsula. As becomes its superpower status, the United States overstraddles all claims with its base at the South Pole. Of the many other nations with bases in place or in plan, their right to be there is "legitimised" by a declared desire to join an international scientific community.

But *realpolitik* demands something more than co-operative international scientific endeavour here in the furthest south. Chile and Argentina have a strong military presence, and other states use civilian contractors for essentially military missions. Antarctic skies are unusually clear and free from radio interference, which makes them highly suitable for deep space research and satellite tracking, and also for operating covert surveillance systems and for the remote control of offensive weapons systems.

Even this might not be enough to explain the underlying reasons for this international presence were it not for an estimated 200 million barrels of oil – more than Kuwait or Abu Dhabi – that are reckoned to lie deep below the Antarctic ice. This must be a strange community then, with competing sovereignty and national resource strategy dressed up as peaceful international co-operation in scientific endeavours.

At one time, these stations comprised not much more than a collection of wooden huts but now they tend towards ultra-modern designs based on appropriate technology to protect the environment whilst coping with the challenge of shifting ice. They have become what one observer describes as "embassies on ice ... showcases for a nation's interest in Antarctica – status symbols". [10]

With territorial claims and related mineral prospecting only temporarily on hold, nations are consolidating their presence by prestigious constructions whereabouts they may be perceived to be engaging in the necessary "substantial research activity" that gives them a vote in meetings to determine the continent's future.

Easter Island, 3,700 kilometres from the city of Valparaíso, was administered by the Chilean military in the recent past, and considered strategically important as a forward base against attack from the west. Its militarisation was also motivated by the psychostrategic anxiety that, as we have seen, often afflicted "island-eating" nations – in this case, Chile was concerned that an historical enemy like Peru could make use of Easter Island if no action were taken. In 1985, a plan was considered to extend the airstrip, ostensibly to provide an emergency landing site for the US space shuttle, but with critics claiming it was an attempt to bring Easter Island within the sphere of the controversial Strategic Defence Initiative. [11]

The South Atlantic's Ascension Island was taken by Britain in 1816 as a precautionary measure when Napoleon was exiled on St Helena 1,100 kilometres away, and for which reason a garrison was temporarily established on Tristan da Cunha, too. Ascension has been used as a sanatorium for yellow fever patients, as a coaling station, and as a stopover during the brief heyday of flying boats.

It was employed as a telegraphic station during World War I, and a naval and air station during World War II. Its runway was once the longest in the world and used to operate the Space Shuttle; NASA tracked moon landings and the European Space Agency has tracked its rockets from the island. There is also a BBC World Service relay station, and it is the site of a joint Royal Air Force and United States Air Force base – where reports suggest the Americans call the shots.

Its military significance was apparent during the Falklands War in 1982, when the 3,000-metre runway provided Britain with a vital support base, and it is easy to understand how the British Foreign Office could perceive Ascension's value as being in its location alone. This is reinforced by the fact that, with a population consisting almost entirely of 700 people from St Helena, 100 British and seventy Americans, Ascension is considered to have no native inhabitants (a claim disputed by island council committee members). It is described as "a working island", the British government has asserted that there is "no right of abode" on the island, and entry is by permission of the Administrator. Only in 1999 was it announced, in a manner confirming that Britain was very much beholden to the United States in the matter, that the two countries were beginning talks about civilian aircraft access, thus opening up its lava moonscape as a tourist destination. There may be increasing pressure to use Ascension for commercial flights since the opening of St Helena's £285.5 million new airport has been suspended indefinitely for commercial traffic (see Chapter Five). Ascension could perhaps provide a refuelling stopover for the smaller-than-planned aircraft that St Helena may eventually accommodate, but there would be strong voices against the failure that this would imply.

With their capacity for existential ambiguity, it is perhaps unsurprising that some islands at various points in their history have been administered by navies, and virtually treated as if they were ships. This has been the case in American Samoa and Guam, Easter Island and the Juan Fernandez Group. "HMS Diamond Rock" was the name applied in Victorian times to one of the Grenadines, in the Windward Islands. In 1942, with reports of a German naval presence in the area, Tristan da Cunha was named "HMS Atlantic Isle" and housed military communication installations. There was even a story circulating about low-lying Inishmurray Island, off the coast of north-west Ireland, concerning a German U-boat in World War I mistaking the island for a ship and firing torpedoes at it. Similarly, Rockall, far off the west coast of Scotland, is said to have been mistaken for an iceberg, a sailing ship, a whale and a submarine. A merchant cruiser during World War I mistook it for an enemy vessel, offered it a chance to surrender, and then opened fire on it. [12]

Because of the strategic importance of small, remote islands, governments will often seek a presence in order to protect their interests during times of unrest and war.

In the Pacific during World War II, even tiny remote isles were briefly dragged out of obscurity to gain an unlikely significance. Radio stations were established on places like Anchorage Island, in Suvarov Atoll, which is about one kilometre long, 300 metres wide and only four metres above sea level at best. The South Pacific atoll is 385 kilometres from Manihiki, and 930 kilometres from Rarotonga, which is the administrative centre for the Cook Islands. But, more significantly, it was never on a shipping route. Indeed, as discussed in Chapter Two, its remoteness was a persuasive factor in its selection by R.D. Frisbie and Tom Neale as an ideal island for voluntary maroonment. Yet, despite its apparent insignificance, Anchorage Island was occupied by New Zealand military personnel during World War II. And there were similar unlikely places elsewhere in the Pacific where US military "were exiled to sit out the war ... on 'lost islands'" far from the combat zone at tiny communication bases, like Tongareva, Aitutaki, and Bora Bora. [13]

In the Atlantic, the Orkneys, Faroes, Iceland and south Greenland were transformed into military bases from which the Allied navies and air forces defended "The Atlantic Trench".

Also during World War II, the Scottish island of Gruinard – about 200 hectares, and three kilometres from the mainland – was requisitioned for biological warfare testing, and in 1942 parts of the island were contaminated with anthrax spores. Gruinard was closed to the public and, set in an area of outstanding natural beauty, it stood as grim evidence of warfare's long reach. (In fact, the closure was difficult to enforce. I sailed there in the mid-1970s, and despite warning signs evenly spaced every few hundred metres, succeeded in wandering about feeling only mildly anxious.)

The British Ministry of Defence (MOD) – master of denial, obfuscation, delay and ignoring requests for information – neglected cleaning up Gruinard even though it was under contract to return the island to the original owners after the war. According to a story circulating locally in the early 2000s, an eco-terrorist group telephoned the Conservative Party Conference in 1985 claiming they had put anthrax spores from Gruinard in the air conditioning system of the hotel where Mrs Thatcher was staying. That this was a hoax was revealed only after a security check; but within months, the MOD announced a clean-up of the island, costing half-a-million pounds, which was carried out the following year. An independent group of scientists declared it safe in 1990. In 1991, disregarding a minority expert opinion that *Bacillus anthracis* can survive for hundreds of years by leaking through subsoil and rocks to form underground pockets, the Scottish Highland Regional Council supported a planning application for a dwelling house to be used for part of the year. The island's owners were given a 150-year indemnity by the MOD against an outbreak of anthrax poisoning. [14]

As an aside, it should be noted that isolating experimenting and testing on islands has widespread appeal. The curious case of Huemul, an island in Argentina's Nahuel Huapi Lake, has an almost surreal touch. Here, in the 1950s, it was reported that an exiled Austrian Nazi used one percent of Argentina's gross national product in building a cold fusion reactor and trying to win the race to "safe" nuclear energy production. He failed, of course. [15]

More significant is the history of Wardang Island, South Australia. In 1859, twenty-four rabbits were imported to Australia and released for sport. Forty years later some 200 million rabbits had virtually destroyed huge swathes of agricultural land and were having a profound effect on ecosystems across the country. Farmers were driven from their land, and some native animals, like the bilby (a desert-dwelling marsupial), were pushed to the verge of extinction.

Myxomatosis is a dreadful disease, but for many years it controlled rabbit numbers until it began losing its effectiveness as populations developed resistance. For some years Australian scientists had been carrying out experiments seeking a replacement. In October 1995, on uninhabited Wardang Island – just four kilometres off the South Australian coast, and where the first field trials for myxomatosis were carried out in 1937 – scientists were testing the lethal effectiveness of rabbit calicivirus disease (RCD). It was believed that a small island would be a highly suitable natural laboratory where security and containment could be assured, but this confidence was shattered when it became apparent that the disease had made the leap onto the mainland, the suspected carrier being a ubiquitous bush fly that can travel hundreds of kilometres. Thousands of wild rabbits were killed on the near mainland in an attempt to control the spread, and the government made a vaccine available to rabbit breeders and domestic pet owners. It is estimated that RCD killed up to ten million rabbits within eight weeks, and by the time the effectiveness of the virus was levelling out, some fifty per cent of the Australian rabbit population had been killed, the figure reaching ninety per cent in some arid areas. [16]

The "anthrax island" of Gruinard may or may not be safe for human habitation, and elsewhere in the British Isles some areas experienced intense militarisation during World War II, which has left long-lasting scars on the landscape. In the Orkney Islands, Scapa Flow was a safe haven for organising Atlantic convoys. And in the Channel Islands, which have a long history of sporadic militarisation, the occupying German forces established defence systems aimed at obstructing British movements to and from Europe.

Decades later the island of Hoy in Orkney could still be described as "like a Yukon shanty town abandoned after a gold rush" where "every field has its military installation, ruined but imperishable; monstrous slabs and blocks of concrete; abandonment and neglect seemingly more acceptable on a small island, out of sight and out of mind from mainland preoccupations". [17]

Likewise, in Alderney in the Channel Islands, Kevin Crossley-Holland describes "ordnance, rampart and contention ... colossal unleavened lumps of brick and cement capping rock crags like helmets; the graceful sweep of a bay destroyed by a massive anti-tank wall." Massive, seemingly irremovable bunkers, derelict camps with concrete bunkhouses once used to house slave labourers give the appearance of an island at the end of the world rather than in the English Channel. [18]

In the Scottish Hebrides, militarisation has taken advantage of the fact that even islands close to shore are relatively difficult to access. In the past they have been requisitioned with limited public reaction, especially if they were unpopulated, and had the advantage that military activities could be easily hidden from close public scrutiny. Strategic military considerations have traditionally been the concern of a government hundreds of kilometres away in London, with input more recently from the devolved government in Edinburgh, which is also hundreds of kilometres away. Feeling marginalised, local people are less than sanguine about how their islands are graced by projects to test rockets, torpedoes and shells that would be unacceptable elsewhere.

And protest abounds. Plans to spend some £40 million on a NATO base on Lewis inspired a petition of more than 40,000 signatures against the proposal. Despite the advantages of employment and infrastructure it would bring, the islanders apparently had no desire to have their relatively insignificant strategic military status transformed into a nuclear target zone. Likewise, the government's identification of six possible sites for the disposal of nuclear waste precipitated a vigorous "Hebrides Against Nuclear Dumping" campaign in the 1980s, an issue that has rumbled on into the 2000s as evidenced by the UK Defence Secretary announcing in 2008 that nuclear waste would not be stored on uninhabited Sandray and Fuday (south of Barra in the Outer Hebrides) if there were local opposition – which there undoubtedly was.

Lewis may have avoided militarisation, but there has long been a missile launching range on South Uist and Benbecula, from which flights are tracked from St Kilda. Even by the mid-1970s, nearly £7 million had been spent on military installations on the Benbecula–South Uist site, and £20 million spent on the Benbecula–St Kilda range. By the end of 1993, the MOD was considering transferring the missile range to the south of England. But what had started out as a

project perceived by the local population with strong suspicion and fear for the future of island culture, was by now a
project perceived – if not by all sections of the population,
then certainly by their parliamentary representatives – as a
fundamental necessity for the existence of the community.
Quoting a plethora of economic statistics, the local member
of parliament stated that running down the military sites
would be a mortal blow from which he could not see the
islands recovering. So it had taken about twenty-five years
for Benbecula–South Uist to become dependent on a
military presence, and a further sixteen years for the MOD
to announce in 2009 that any plans to close the range had
been scrapped. [19]

St Kilda has also been militarised as part of this programme,
with a missile tracking station being built during the late
1950s at a cost of nearly £1 million. In 1969, a village was
constructed for service personnel, meaning that St Kilda is
now home for a transient workforce. Ironically, the island
group is owned by the National Trust for Scotland – holding
the island in trust for the nation as a whole – and managed
in collaboration with Scottish Natural Heritage, dedicated
to countryside conservation whilst stressing the value of
human society within that countryside. It was Scotland's
first World Heritage Site, recognised for both its cultural
heritage and it marine environment.

Nuclear Testing Sites

The most catastrophic effects of militarisation are evident
on Pacific islands. Oceania, comprising Polynesia, Melanesia
and Micronesia, is made up of about 3,000 tiny islands with
a combined population of about 1.5 million. This sprawling

area had been economically and politically self-sufficient in some instances for as long as 4,000 years. Lagoon fishing and small-scale agriculture made survival relatively easy, factors that prompted early Western voyagers to describe a vision of paradise.

In 1935, President Roosevelt placed Wake Island under the administration of the US Navy and granted Pan American Airways permission to build runways on Wake, Midway and Guam. Considered a clear provocation by Japan, a retired commander-in-chief of the Japanese navy described these islands as "natural aircraft carriers", affording enemy squadrons with ideal locations from which to operate, and from which they could seriously endanger Japanese defences. As events turned out, the military value of Pacific islands during World War II led to island-to-island combat between US and Japanese forces, ravaging the Gilberts, the Carolines, the Marianas, and Fiji in particular. [20]

Later, the United States began nuclear testing in the Marshall Islands, whilst other nations, most notably France and Britain, used their own "possessions" for the same purpose. Micronesia was awarded to the United States in 1947 as a United Nations Strategic Trust Area. In return for $100 million a year and access for its citizens to work in the United States, islanders accepted that their ancestral homes were now designated "test sites" and the recipients of billions of tons of toxic explosives. Identified by one critic as "the worst example of recent US behaviour towards a native society", the recent history of Micronesia has been reduced to a grim catalogue of events, including the trauma of people being uprooted from their homelands and being subjected for generations to radioactivity-related diseases.

This has only been possible on remote, small islands isolated from continental view. [21]

In the Marshall Islands, the people of Bikini were evacuated in 1946 to Rogerik Atoll, 200 kilometres to the east. This was to be temporary and "in the cause of peace". The islanders discovered poor fishing, insufficient fruit and coconuts and few government services, and the Bikini islanders requested they be returned home. But by then nuclear explosions had made their home uninhabitable. So in 1948 they were relocated on Kwajalein Atoll, 300 kilometres further to the south and the target of missiles fired from Vandenberg Air Force Base 7,500 kilometres away in California. Later that year, they were moved again, this time to Kili Island, 400 kilometres to the south-east. Some of these "nuclear nomads" returned to Bikini in 1971, but by 1978 they were found to have ingested unsafe levels of Celsium 137, and the island was evacuated again.

Bikini was used for twenty-three tests in the 1940s and 1950s, and was selected because US personnel believed they could predict wind direction around the atoll. However, unanticipated wind shifts during the test period are estimated to have spread material over an area perhaps as large as 80,000 square kilometres, seriously contaminating Rogerik and Rongelap Atolls. On Rongelap, three out of four children under the age of ten developed thyroid tumours after the "Bravo" test in 1954. And some were evacuated to Ejit Island on Majuro Atoll. The people of Eniwetok were moved to Kwajalein even though it had been found to be unsuitable for Bikini refugees. On Kwajalein, 8,000 people were evacuated onto the twenty-eight-hectare Ebeye Island to help in the construction of a base. With no lagoon, no fishing and no room for agriculture, the population is twelve times the density of an urban area like Washington DC. It has been claimed that those not evacuated because of their labouring skills lived in "Tintown" and, "lest they spread secrets, they are not allowed to leave Kwajalein and other Micronesians may not enter". [22]

No matter how one views the role of nuclear weapons in the world, it is difficult to apprehend how the US government could get away with exploiting vulnerable people in such a cynical way. The suffering of these people, the destruction of culture and of economic livelihood, is appalling. No amount of compensation can reverse the catastrophic consequences of dislocation. In a Compact of Free Association with the United States, Bikinians have received more than $20 million for resettlement, but are claiming $450 million in court for the deprivation of and permanent damage to their island home. The journalist William Ellis and photographer James Blair recorded the anguish expressed by Bikinians, about half of which – roughly 650 – live on Kili. They describe a people without social autonomy, as the old yearn for a return to Bikini and the young seek the opportunity of yet another resettlement, this time in Hawaii. They have virtually lost their skills as fishermen and sailors, and "as wards of the US government they (especially those born after the 1946 evacuation) have become addicted to welfare ... with nothing to do but watch for the supply plane or dream of being somewhere else." [23]

The United States is spending $90 million on removing topsoil and "decontaminating" Bikini's 250 hectares, but it has been estimated that without applying enormous amounts of potassium-rich fertiliser to the soil, it could be seventy years before the atoll is safe for crop production. On Runit Island in Eniwetok Atoll, the problem of contaminated land has been tackled by burying some 100,000 cubic metres of radioactive soil beneath a half-metre thick dome of concrete. Elsewhere in the Marshall Islands there are similarly serious issues of contamination on Jaluit and Rongelap Atolls.

During the 1980s, whilst undertaking research and development for the Strategic Defence Initiative and creating monitoring systems for a new global navigation system for ships and submarines, United States spending on Kwajalein was $40 million. This suggests that militarisation of some variety will continue to be very significant on the atoll. As an adjunct to this it should be noted that in 1947, under the United Nations Strategic Trust Territory mandate, the United States formally agreed to promote the health and economic self-sufficiency of the inhabitants, to regulate and control the use of natural resources, to protect the inhabitants against the loss of their lands, and to improve transportation and education. They were also mandated to give up trusteeship within twenty-five years – that is, by 1972 – and to permit self-governance under terms dictated by the people. [24]

France, having lost test sites when Algeria gained independence in 1962, has undertaken nuclear testing on Mururoa Atoll and Fangataufa Atoll, in the Taumotu Archipelago of French Polynesia. Between 1966 and 1992, they undertook forty-one atmospheric tests and 138 underground tests. These ceased in 1992 due to Non-Proliferation Treaty obligations. Then, in June 1995, amidst considerable international consternation, President Jacques Chirac lifted the moratorium and continued underground testing – launching eight tests between September 1995 and May 1996, with the justification being that France wanted to create computer simulations that would eradicate the need for future testing.

Not for the first time, this activity provoked protests in Papeete, where the international airport was among buildings damaged by rioting. The French embassy in Australia was burnt down by the Pacific Popular Front. Crumbs of consolation were tossed to the media when journalists were provided a rare opportunity to visit the test site on Mururoa, described by one in a piece titled "Paradise Lost to Nuclear Mayhem", as "a curious cross between a holiday camp and a laboratory for mass destruction". [25] Despite the atoll sinking to less than three metres above sea level as a result of many years of underground explosions, despite the construction of a large concrete wall to keep the sea out, and despite thousands of tons of concrete being used to slab over explosion chambers, one of French Polynesia's representatives in the French Parliament commented that, when the nuclear tests ended, "Mururoa can look forward to a new future as a holiday resort". [26] France is now co-signee of a new moratorium against nuclear testing.

As for the military personnel who took part in nuclear tests, the United States, France, China, Canada and New Zealand have paid varying levels of compensation to their citizens. About 28,000 service personnel were involved in twenty-one atmospheric tests carried out by Britain on mainland Australia and at Malden Island and Christmas Island between 1952 and 1958. With regard to the 1950s

Christmas Island bomb tests, even the British MOD admitted in 2009 that ninety per cent of 114 "essential witnesses" were dead. When "lead cases" representing a total of 1,011 cases (with an average age of nearly eighty-four) were submitted for legal redress, a High Court judge decided that the claims should go ahead. Subsequently, the Court of Appeal and the Supreme Court judged that it had been too long since the problems emerged to assess cause. Just one "lead case" was continuing, which might allow other veterans to seek compensation.

And so we are left with the recollections of these elderly veterans. Douglas Hern served in the Royal Navy and describes how at dawn on April 28, 1958, he and his comrades were marched down to the beach on Christmas Island and ordered to change into blue overalls, anti-flash gloves and balaclavas. They were told to sit down on the beach, with their hands over their closed eyes and knees drawn up, as a countdown from ten began. When the bomb exploded, "there was a feeling of intense heat. We were told to stand up and look. We saw a bright, brilliant light. It was as if someone had switched a firebar on in your head. It grew brighter and you could see the bones in your hands, like pink X-rays, in front of your closed eyes. We saw a swirling mass of orange and blue and black and red and the noise was like fifteen underground trains coming at you." [27]

It is perhaps unsurprising, then, that the people of Palau – comprising about 250 islands with a population of 21,000 people in western Pacific Micronesia – voted regularly to reject a Compact of Free Association with the Unites States, despite having been offered significant "financial incentives". Against considerable pressure, Palau maintained a Nuclear Free Zone, thereby denying itself very significant aid money. Rumours abounded of bribery, political skulduggery, and even of CIA involvement in political assassination. The US interest is in the use of islands for jungle warfare training, airfields and weapons storage. Most importantly, Palau has considerable strategic importance as a potential deep-water harbour for nuclear submarines. In 1993, after eight referendums, the United States was successful in "changing minds", and Palau received a first instalment of $250 million in return.

The Case of the Chagos Islands

The recent history of the Chagos Islands – a far-flung cluster occupying about 54,000 square kilometres in the Indian Ocean – illuminates many of the issues outlined in this chapter. The largest island is Diego Garcia, a horseshoe-shaped lagoon about twenty-two kilometres by six kilometres. The substantially smaller Peros Banhos and Salomon groups, comprising some thirty-five outer isles, lie 300 kilometres to the north of Diego Garcia.

The Chagos are noted for their high species diversity and unique marine habitats, described by environmentalist and journalist Michael McCarthy as "among the least polluted marine locations on Earth. Its seawater is the cleanest ever tested; its coral reefs are completely unspoiled; its whole ecosystem, with its countless seabirds, turtles, coconut-cracking crabs (the world's largest), dolphins, sharks and nearly 1,000 other species of fish, is pristine." [28]

Human settlement is first recorded in the mid-1780s, with a coconut plantation under French rule worked by Malagasay and Mozambican slaves. At about the same time the islands housed a leper colony for Mauritius, occupied by 300 patients. After the defeat of Napoleon, Britain took possession of Chagos in 1828 with a population of 448, more than half on Diego Garcia and with settlements extending to the outer isles. Indentured labourers were introduced from India in the 1840s and 1850s.

By 1900, the population had risen to about 750, with about 500 on Diego Garcia, where guano was exported for phosphate fertiliser, three small copra factories operated and there was a church, hospital and coaling station. Most families fished, had small vegetable gardens and kept chickens and ducks. A film made in the 1950s by the colonial authority noted people who "lived their lives in surroundings of wonderful natural beauty and in conditions most tranquil and benign". [29]

Thus, it appeared that, at least in this remote place, there were some good things in the twilight years of the British Empire. Indeed, a visitor reported in the late 1950s that, "It was paradise there. ... There was a *château* ... whitewashed stores, factories and workshops, shingled and thatched cottages clustered around the green ... lamp standards and parked motor launches." [30] Reports from visitors and from the French-run copra and coconut oil company reinforce the perception that the Chagos Islands supported a small, economically viable and generally contented society.

Ten years later all this had changed most tragically. By the 1960s, Britain and the United States were reviewing their sphere of influence and looking for new sites for military bases. Increasing aspirations for independence meant that the future of operations in, for example, the Maldives or Singapore, were uncertain. There was war in the Yemen, and guerrilla warfare in Oman where oil was discovered in 1964. These events threatened access to the Persian Gulf and the Red Sea; and perhaps too the future security of oil supplies from the Middle East was anticipated as more important than the vague threat posed by the Soviet Union in the area.

The first choice of the United States for a base was in the Aldabra Atoll in the Seychelles. But this proposal was actively resisted by formidable establishment bodies like the Smithsonian Institute in Washington and the Royal Society in London on account of its significant ecosystems. But for reasons unclear, it appears that neither its "pristine ecosystem" nor its human settlements living in surroundings "most tranquil and benign" were considered reason enough to summon similar forces to protect the future of the Chagos. [31]

Early in 1964, the British government negotiated the independence of Mauritius – until then a colony of which the Chagos Islands were a part. But for £3 million and the promise of a sugar import preference, Chagos was to remain British. To do this the government was obliged to disregard a United Nations resolution calling on them not to violate the territorial integrity of Mauritius by separating Chagos at independence. This was achieved by the Queen approving an Order-in-Council that bypassed Parliament, and which presented the United Nations with a *fait accompli*. Chagos became part of the British Indian Ocean Territories and, through negotiations carried out with the United States *before* the independence of Mauritius, was leased to them as a military base. [32]

For a fifty-year lease, negotiable for a further twenty years, it was reported the UK received a discount of $14 million off the price of Polaris submarines. For the islanders, the price was much higher: declassified documents indicate the deal, described by negotiators at the time as a "neat, sensible package", involved the archipelago being "swept clean" and "sanitised". [33] About 2,000 people, who could trace their lineage through several generations on the islands, were to be forcibly evicted as security risks. According to Article 7(d) of the Rome Statute of the International Criminal Court, which established the International Criminal Court (ICC), "deportation or forcible transfer of population" constitutes a crime against humanity if it is "committed as part of a widespread or systematic attack directed against any civilian population, with knowledge of the attack". The British government could probably have argued its way around this, but it was secure in the knowledge that the ICC is not retroactive, and it is unable to consider alleged crimes committed before July 1, 2002. [34]

The very least a colonising power is responsible for is the fair administration of indigenous populations. To collude in their deportation is as much a gross violation of human rights as it is immoral. But the British government sought to avoid conflict with international law by claiming that there was no indigenous population, and that the inhabitants were merely transient labourers on a copra plantation.

The deportations, started in 1967, were complete by mid-1971. In a series of diplomatic manoeuvres the islanders' culture and economy was torn asunder. They were stuffed into ships and exiled far away in Mauritius and the Seychelles.

One resident expressed the trauma of events, of being uprooted from where she and three generations of her family were born and where she had mothered six children: "What

I can't forget is the fear and uncertainty for myself and my family. When we got to the Seychelles, the police were waiting for us. They marched us up the hill to a prison, where we were kept in cells until the boat was ready to take us on to Mauritius. I suppose we took some hope in the promise that in Mauritius we would be granted a house, a piece of land, animals and a sum of money. We got nothing." The former president of Mauritius takes up the story, describing the people as "bewildered" and "terrified ... who would sing their way through life; and here they were, weeping their way through life, and they are still weeping". [35]

The legacy of the deportation is of Chagossians living on the margins of Mauritius society, with poor living conditions, high illiteracy, high unemployment, high suicide rates and poor health, with young people increasingly turning to drug and alcohol abuse and prostitution. It is claimed that some Chagossians were living twenty-five in a room and in shifts, compensated by the British government to the extent of about £650 per adult and £3 per child. Some 422 families signed a petition indicating they wished to return home, and eventually the government offered a further £1.25 million compensation – if islanders signed a "no return" clause. It was rejected, and women protested in the street as these dreadful conditions were allowed to deteriorate in Mauritius. With pressure brought to bear by the Mauritian government, the UK government responded by cancelling a job creation scheme (unemployment among Chagossians was running at sixty per cent), and using the money to offer the equivalent of £2,000 to each islander.

In November 2000, a High Court ruling and government ordinance allowed islanders to return to Chagos, but with neither compensation nor assistance, and with a stipulation that they could not return to the main island of Diego Garcia. Return under these conditions was impossible for almost all. In June 2002, the Foreign Office produced a report arguing the excessive cost of trying to repopulate the islands in the long term; and in any case, in June 2004, the Foreign Office overturned the 2000 court ruling and ordinance. In May 2006, the High Court found the UK government's action illegal, and at the subsequent Court of Appeal in May 2007 it was ruled that the islanders could return. The Chagossians indicated that some 5,000 people wanted to return, and that this was something the government should pay for.

An editorial in *The Independent* in May 2007, the day after the Court of Appeal ruling, describes the expulsion of the Chagos islanders as "an outstandingly shameful episode in post-war British history". And it goes on to assert that, "looking back, it is quite astonishing to imagine what happened. ... It was reminiscent of one of Stalin's deportations of native peoples." The Court of Appeal refused to allow the British government an immediate right of appeal, but the Foreign Office was considering a petition to the House of Lords directly to review the case, a fact that outraged many observers.

Outrage or not, the Foreign Office did indeed petition the House of Lords. And by a majority of two out of three their lordships upheld the government, citing security issues and the costs of re-establishing a permanent settlement.

In a final twist to this sorry tale of Machiavellian proportions, the Foreign Office announced plans to establish what was then the biggest marine nature reserve in the world centred on the Chagos Islands. Without mentioning the Chagossians directly, the 2010 announcement was made "without prejudice to the outcome of the current pending proceedings". This has led the co-ordinator of the Chagos Islands All-Party Parliamentary Group to declare that it is as if the Chagossians have been "airbrushed out of the press release, and out of existence". [36] The Chagossians would probably welcome a marine protected area, but only one in which their rights of return were uncompromised and in which they were fully integrated as contributing members.

But it would appear that their rights of return may forever remain unrecognised. In December 2012, the European Court of Human Rights in Strasbourg ruled that the islanders had already accepted compensation, so no further compensation could be considered. And in November 2016, the UK government confirmed there would be no return, citing feasibility, defence and security, and cost. The £40 million of compensation would be spread over ten years. At the same time it was announced that the United States' lease on its base on Diego Garcia would be extended for twenty years.

For the moment, the Chagossians are barred from returning, but the matter still has its frayed edges. In the immediate aftermath of the announcement that there would be no return for the islanders the prime minister of Mauritius, Sir Anerood Jugnauth, accused the UK of undermining human rights on the Chagos which, he said, "always formed and continues to form an integral part of the territory of Mauritius". He claimed, too, that Mauritius would be fully justified in taking the issue of decolonisation (which the UK had frustrated by separating Chagos from Mauritius) to the UN General Assembly with a view to putting the matter before the International Court of Justice at The Hague.

The UK government denies the Mauritius claim of

sovereignty and reasserts its undertaking to cede the archipelago to Mauritius when Diego Garcia is no longer required for military purposes. [37]

It is difficult to envisage the rationale for a huge internationally significant nature reserve coexisting with a colossal military base and its personnel whilst arguing that a few hundred Chagossians would have an unacceptably detrimental effect. However, reported diplomatic leaks suggest that the Foreign Office had indeed told US representatives that the marine reserve would effectively end the islanders' resettlement claims. [38] Of course, the military base could be excluded from the reserve, but this would significantly fragment the integrity of the island ecosystems and set the plan for a reserve in direct conflict with basic conservation principles.

According to the United States, by the mid-2000s, Diego Garcia had seen the most dramatic build-up of any base since the Vietnam era. This was driven by an Air Force White Paper establishing the base as the mainstay of their bomber strategy for the next century. With its 2.5 kilometres of runway, and over $500 million spent on construction and infrastructure, it has been used to strike directly and indirectly in Iraq, Somalia and Afghanistan. In 2008, after repeated denials, the British Foreign Office admitted that the public had been "misinformed" and that rendition flights had refuelled there in 2002. Suspicion remains that the base was used for torture, too, a suspicion that became front-page news in July 2014 when the Foreign Office claimed that key documents had been lost due to "water damage". For Chagossians it seems that paradise too has been lost – lost to make way for what the United States calls, without irony, "Camp Justice".

Chapter Seven

Paradise and Purgatory

The Island

Adrian:	The air breathes upon us here most sweetly.
Sebastian:	As if it had lungs, and rotten ones.
Antonio:	Or as 'twere perfum'd by a fen.
Gonzalo:	Here is everything advantageous for life.
Antonio:	True: save the means to live.
Sebastian:	Of that there's none, or little.
Gonzalo:	How lush and lusty the grass looks! How green!

William Shakespeare, *The Tempest, Act II, Scene I*

According to the Bible, man and woman lived in Paradise until they fell from grace for the sin of disregarding the will of God. For many millions of people, achieving God's forgiveness for events in the Garden of Eden enables them to anticipate a time when they will ascend to a heavenly paradise. But this paradise has not always been located only in a celestial heaven. Some, like the Romans and the Greeks, did not believe in the concept of original sin, but were understandably attracted by the idea of a paradise, an Earthly paradise located in a world so much of which was still an unexplored mystery.

The widespread appeal of a paradise to be found on Earth stimulated voyages, idealised by Homer, in search of what were variously named Arcady, the Gardens of the Hesperides, the Elysian Fields and the Islands of the Blessed. The Celts of Wales later added Avalon, and the Irish the Land of Youth and the Fields of Happiness. Here was to be found, in the words of the Isaac Watts hymn, "a land of pure delight where saints immortal reign ...where everlasting spring abides and never fading flowers".

The search for this paradise was inspired first by accounts

that are mostly mythological in origin, but with the passage of time events gained impetus as myth became legend which became stories of uncertain veracity. This process involved deciphering a tortuous mix of fact and fiction accumulated from voyages of discovery.

The Greek poet Hesiod (ca. eighth century BC) described a race of hero men who were presented with an Island of the Blessed by Zeus, where they lived untouched by trial or sorrow, whilst the Roman Horace, despairing of the Republic in a time of civil war, urged his compatriots to voyage "to these fortunate islands, where the land without tillage produces grain abundantly, and the vine bears grapes without pruning". [1]

But voyaging to these islands was no easy matter, as they were usually located beyond the Pillars of Hercules, west beyond the Mediterranean and the known world, where heroes beloved of the gods were rewarded with immortality. So this place was worth a lot of trouble looking for. The islands' enchantedness is captured by Plutarch (46–120 AD) in describing the Fortunate Isles: "Rain seldom falls there, and then only moderately, while they have usually soft breezes which scatter such rich dews that the soil is not only good for sowing and planting, but spontaneously produces the most excellent fruits; and these in such abundance that the inhabitants have only to indulge themselves in the enjoyment of ease and leisure ... Hence it is generally believed, even among the barbarians, that these are the Elysian Fields, and the seats of the blessed, which Homer has described in all charms of verse." [2]

Avalon is embedded in Welsh Celtic tradition, an Elysian island retreat in the west to which King Arthur was to be borne and healed of his wounds after the Battle of Camlann.

The Island

Avalon is, according to Arthurian legend, "a green and fertile island which each year is blessed with two autumns, two springs, two gatherings of fruit – the land where pearls are found, where the flowers spring as you gather them – that isle of orchards called 'Isle of the Blessed'." [3]

This tradition of an enisled earthly paradise is not an exclusively Western phenomenon. The geographer Y-Fu Tuan has emphasised the symbolic significance of islands in Eastern cultures, where they represent "a state of prelapsarian innocence and bliss, quarantined by the sea from the ills of the continent." [4] According to Tuan, Buddhist cosmology is focused upon four islands, just as Hindu doctrine features a significant island, with many gems and sweet-smelling trees. For the Semang and Sakai of Malaysia, paradise is located on a fertile island; and it is probable that this belief in Elysian islands influenced Marco Polo's thirteenth-century description of the islands of Male and Female – which he seems not to have visited – located about 500 miles south of India, where the climate would be mild, the land fertile, and the people led simple, contented lives. [5]

Between 350 and 250 BC, ships began to sail forth from the north-east coast of China in search of the Islands of the Blessed, hoping to find the mushroom that guaranteed immortality. Taoists were seeking the elusive elixir of life, a search in which Lieh Tzu spoke allegorically of the island of P'eng-lai in an archipelago far out to sea where white men lived in buildings of gold. "Immortal sages live there, who eat sweet flowers and never die." [6]

By the late fifteenth century, geographical knowledge was very slowly disentangling itself from mythology, but the belief in an enisled paradise was an *idée fixe* that could not easily be dispelled. Portugal alone, for example, sent out at least eight expeditions between 1462 and 1487 with the primary objective of searching for new Atlantic islands, focusing especially on the islands of Antilia and Seven Cities. Although often located west of the Azores on fifteenth-century charts, considerable confusion arises in trying to fit these islands onto a modern chart. Antilia has been plotted as far away as Cuba, presumably in recognition of the fact that Columbus believed he had discovered Antilia, which he anticipated using as a way station en route to the "Indies". And confusion also exists with Seven Cities, which has been equated with Antilia, but which perhaps has a history dating back to as far as 711 AD when seven bishops, fleeing from the Moors, reached Portugal and took ship to settle on a great island to the west where they built seven cities. [7]

Paradise Found

By the mid-eighteenth century, seafaring European nations had a more coherent picture of world geography, and paradise still remained an elusive goal. The Caribbean, North American, African and "East Indian" natives were not perceived as living in the Garden of Eden, and were promptly earmarked for commercial exploitation, slavery or extinction. Indeed, the "colonial experience" seemed to support the mordant view of the philosopher Thomas Hobbes (1588–1679) that, devoid of the trappings of "civilisation", life was solitary, mean, brutish and short.

It is fortunate, therefore, that Louis-Antoine de Bougainville discovered Tahiti at a time when Europeans were seriously questioning the working of human nature and

the values of civilisation. Philibert Commerson – Bougainville's naturalist in 1769 – described people in Tahiti living
in a fair climate, at one with their landscape, and on land
fertile without cultivation. Indeed, for him it appeared that
"the condition of natural man, born essentially good, free
from all preconception, and following without diffidence
or remorse the sweet impulses of an instinct always sound,
because it has not yet degenerated into reason". [8] Commerson is thus reflecting his contemporary Jean-Jacques
Rousseau (1712–1778), who proposed that over centuries
animal instincts were transcended, and mankind lived both
morally and happily until corrupted by decadent civilisation.

Bougainville called Tahiti "New Cythera", after the
Greek island where legend recounts that Aphrodite, goddess
of love, rose from the waters. And even his pragmatically
inclined European mariners relayed an image of the perfect
tropical island, with a gentle climate where food was so easy
to find, where the burdens of life were light, and where
there was plenty of time for love-making.

Herman Melville, too, who deserted from a whaler in
1842, discovered in the Marquesas a culture he perceived as
being at peace with itself, getting along very well without
the trappings of Western society and its representatives –
missionaries, traders, administrators and so on. And indeed
Melville even went so far as to suggest that Polynesians
should go as missionaries to "civilised" countries! Even a
hundred years after Bougainville, Pierre Loti was able to
talk about "the perfect primitive dream made flesh", a sentiment echoed by the French Ministry of Colonies, which
described a Tahiti with "no winter and no discontent, ripe

fruit for the picking, a workless world where to live was to sing and to love". [9]

One may ponder why some of these travellers were not tempted to remain in their new-found paradises. They were perhaps a little perplexed that their own notion of civilisation was at least loosely based around a work ethic, whilst the indigenes of Polynesia appeared to do only as much work as was absolutely necessary. Many missionaries came not only armed with the gloom and doom of the Old Testament, but also with a predisposition to dismiss the value of play and love-making as devilish distractions. Many of the early traveller-artists-writers were in fact transients – for example Herman Melville was a sailor for only four years, and after the age of twenty-five never worked at sea again. And others held on to their outsider perceptions with a tenacious grip, like Robert Louis Stevenson, who wrote to J.M. Barrie: "I live here in the South Seas under conditions so new and so striking, and yet my imagination so continually inhabits that cold old huddle of grey hills from which we come." [10]

Twenty years later, when Paul Gaugin arrived in Papeete, he described a grubby little port town, which, to his dismay, civilisation had been to grips with for over a century. But he brought with him his overweening artistic ambition that no paradise could assuage. And he died with it – from poverty, a bad heart and failing sight certainly, and with various unpleasant addictions and diseases perhaps. Surely this was no utopia for him.

It seems then that artists and writers have a rather ambiguous attitude to their perceived "paradise"; and the sometimes close proximity between paradise and purgatory is exemplified in the story of the wreck of the *Stirling Castle*. Sailing from Sydney to Singapore in 1836, the brig foundered on a reef on the north-east coast of Australia. Some of the survivors, including the captain and his wife Eliza Anne Fraser, reached Great Sandy Island (now Fraser Island) in a longboat, where they were captured by Aboriginals, who divided the men among family groups to assist in communal work. Mrs Fraser was forced to do menial tasks and considered herself a slave. Her husband died from spear wounds. These are the bare bones of the story Mrs Fraser brought back to "civilisation".

A convict absconder played some part in her eventual rescue, and highly exaggerated versions of her experiences featured in Sydney newpapers, complete with tales of massacres, murder, slavery and various forms of abuse. Mrs Fraser seems to have done little to correct these accounts, and subsequent re-tellings of her ordeal had negative effects on Aborigine–settler relations.

Controversy and a sense of misrepresentation meant that in the twentieth century descendants of the Aborigines who had attempted to help the castaways forced a re-examination of her story. Mrs Fraser had created a small sensation in her tale of bad treatment, but the local people passed down a different story "of white men who survived and were welcomed at first as reincarnated spirits and a white woman who was absorbed into the life of the tribe". The key point here is that the novelist Patrick White was sympathetic to this native account and, as his biographer David Marr explains: "Mrs Fraser had brought back reports of hell, but White [in his book *A Fringe of Leaves*] was putting together the picture of a savage paradise." [11]

It would be too hasty to write off the pervading image of an earthly paradise in the Pacific as a short-lived aberration. At a fundamental, existential level it is easy to imagine landing on a "desert island" and experiencing, through all the senses, its immediate appeal. And this is not merely the domain of those we may suspect of a natural tendency towards dreamy "lotus-eating" (in Greek mythology, lotus-eaters lived a life of peaceful apathy). The sensual appeal of small Pacific islands is illustrated in some unexpected places. In 1869, aged twelve, Frank Bullen went to sea and served in varying capacities for some fourteen years. In 1898, he published an account of a three-year trip around the globe hunting for sperm whales, and written "from the seaman's standpoint". Towards the end of his account he describes the arduous routine of duties that prevented the crew from "degenerating into lotus-eaters", and makes it clear too that even then dreamy thoughts had no place in describing islands in the New Hebrides and Tahiti "and kindred spots with all their savage, bestial orgies". But he goes on to describe his time on "Vau Vau" (Vava'u in Tonga) as "those lovely days spent in softly gliding over the calm, azure depths, bathed in golden sunlight, gazing dreamily down at the indescribable beauties of the living reefs, feasting daintily on the abundance of never-cloying fruit, amid scenes of delight hardly to be imagined by the cramped mind of the town dweller; islands, air, and sea all shimmering in an enchanted haze, and silence scarcely broken by the tender ripple of the gently-parted waters before the boat's steady keel." [12]

It is perhaps strange to turn to a whaler for an image of paradise, so we will turn to a scientist-adventurer for another perspective. Thor Heyerdahl describes the end of an epic Pacific crossing in August 1947 when he and his crew washed up on a small, uninhabited island in the Tuamotu Group of French Polynesia. This was a "heavenly" palm island with its surrounding reef the "gates of paradise"; the "virgin" sand beach was touched only by their footprints and led up to luxuriant bushes crowned with brilliant-white blossoms "which smelt so sweet and seductive that I felt quite faint. ... I was completely overwhelmed. I sank down on my knees and thrust my fingers deep down into the dry warm sand. ...We poured down our throats the most delicious refreshing drink in the world – sweet, cold milk from young seedless palm fruit." [13] All the senses are stimulated, almost simultaneously and almost overwhelmingly: the sight of the island, the feeling of sand between the toes, the smell of the blossoms, the taste of the palm fruit milk, and the sound of waves breaking on the surrounding reef.

Heyerdahl and his crew were intrepid ocean voyagers and rational scientists, too. Their emotions were cathartic – relief at the end of a long and dangerous journey. But there was more than catharsis here: they were in awe of the sensuousness of their new environment. Bengt Danielsson later wrote that his first meeting with a South Seas island led to him being overwhelmed by the colour of the coral, the roar of the surf, the sun drifting through swaying palm trees and by a sense of peace amongst "sun-steeped sands ... [and] the crystal-clear water of the lagoon ... [we] danced and sang with merry, friendly Polynesians who seemed without a trouble in the world". [14]

And so he promised to return, something the Chief of the Raroia cared to remind him of in a series of letters.

Danielsson, a man of his word, was already organising a research expedition in French Oceania, but delayed things in order to spend more than a year with his wife on Raroia, in the Tuamotu Group. They shared the lifeworld of islanders to explore "in what respects it was better or worse than civilised everyday life".

It is worth spending some time with the Danielssons on Raroia, where they arrived in April 1949, because of their insights into what paradise might really mean, the extent to which living on an island can determine or influence human happiness, and the extent that this may be shared by outsiders. The author was an anthropologist who was not seeking evidence of an earthly paradise (a state of bliss that never existed even in the isolated Tuamotus, which even in the "old days" were ruled through often violent inter-clan warfare). He was more interested in describing people who, on the surface, live an unselfconscious, carefree and happy existence, and who do this by behaving instinctively and intuitively rather than by following rational cause and effect modes of thinking and behaving.

The Tuamotu Archipelago is composed of very small, infertile atolls, surrounded by complex and changing reef systems that are extremely dangerous to navigate except by people with very special local knowledge. Unlike much of Polynesia, there was little to attract outsiders to the idea of settling in the islands, so they remained far from shipping routes. By the middle of the twentieth century, they had survived with little attention from explorers, traders, whalers, missionaries and administrators. The French established a protectorate over the islands in the middle of the nineteenth century, but apart from an island-elected chief, a

Papeete-nominated civil official and an administrator for the islands who visited once a year to listen to complaints, the administration seemed to function well through benevolent negligence. There were no taxes and there was no military conscription. The administration discouraged foreign plantation ownership, so all the inhabitants of Raroia – about twenty square kilometres and with a population of about 120 at the time of the Danielssons' stay – had enough land to at least build a house and to supply their need for coconuts.

Here was an island society lacking in differentiation – no group, party or professional distinctions whatsoever – and which had only one way of looking at existence, no religious doubts (they were nominally Catholics), and who never came into contact with other people and races. There were no social classes, there was a uniform system of social conventions, and there was economic equality because they were all equally good at fishing, making copra and performing various crafts. The island community had "succeeded in preserving the homogeneity which is so characteristic of most so-called primitive peoples... [This made it] whole and believing where we are sceptical and relativist."

The island environment provided a warm climate that meant housing was easily provided through basic materials like palm leaves, galvanised iron, imported fir and packing cases. There was clean water in the lagoon for washing, and fish were very plentiful. Rainfall was sufficient for drinking, and the remarkable coconut palm fulfilled a wide range of uses – the root fibres were used for skirts, the trunk for house posts and furniture, the leaves for multi-purpose mats and baskets, and the nut was used for drinking, for making a sauce that could be served with just about anything, and for fermenting beer. The shells could be made into scoops and bowls, and could be made into charcoal. In fifteen minutes enough of the fibre that covers the outside of the shell could be collected to provide fuel for a week.

Time-honoured tradition dictated how much work was done, and the notion of having goals to progressively enhance "the quality of life" was unknown. So a family could quite easily make three tons of copra a month, but made only enough to get by. The trees, which were easily established, would bear fruit after five or six years and were productive for fifty to eighty years, were left more or less untended, as was tradition. This meant that a large proportion of the nuts rotted in the surrounding undergrowth, and a similar proportion was eaten by rats living in the trees. These were problems that could be relatively easily solved, but they were not "problems" for the islanders, who could see little point in working harder or employing labour in order to produce more than immediate needs.

Capitalism, with its emphasis on competition, growth and accumulation, had no place here, something emphasised in so many ways. It seemed that the production of copra even went down as the price went up; and he noted too that, despite a reluctance to do more work than was necessary, annual earnings were high, much of which went to the two merchants on the island who charged exorbitant prices. The islanders, who never knew whether they were in credit or debit to the merchants, knew that prices were high, but were happy to ignore it because that is what everybody had always done, and finding an alternative seemed like unnecessary work.

The only exception to this otherwise uncompetitive economy seems to have been the habit of copying each other in buying senseless or useless things – like bicycles when there were no roads, furniture when there was no room for it in their houses, and appliances that needed power they did not have. In this they could appear impulsive, even reckless; but it seemed that the islanders could do without these benefits of modern capitalism, and even do it joyfully. So in the economic depression of the early 1930s, food items that they had become accustomed to – flour, coffee, biscuits and beef – became unavailable. The islanders, with apparent equanimity, fell back on the "old ways" of eating fish, sea birds and their eggs, roots, and coconuts; and they did not even have to make copra.

Island life could easily be described as carefree, even idle in its true meaning – stripped of work ethic. "Carefree" and "work ethic": for an anthropologist like Danielsson the challenge was to uncouple Western world associations in observing life on "uncivilised" Raroia to ascertain "in what respects it was better or worse than civilised everyday life". Indeed, seeking to bracket out our acculturated sense of values throws into relief the fact that there will always be aspects of "Paradise in the Pacific" which people from the other side of the world could easily do without. Even as recently as the early twentieth century, diet would have consisted of raw fish and mussels, sea birds, turtles, coconuts, pandanus fruit, taro and pokea (a type of purslane), and products derived from the coconut. Drink would have been water and coconut milk. By the middle of the century, no taro or pokea was grown, and little raw fish was eaten. Whereas the "primitive" diet was repetitive, the modern one was equally so in that tinned beef and dumplings was the outstanding dish of choice, on the basis that it was easy to prepare. There were neither dairy products nor green vegetables consumed, and stomach and tooth problems were endemic.

Danielsson also describes men, women and young people of both sexes as being, without exception, chain smokers; and this hazard coexisted with widespread alcoholism, with islanders being careless even to the extent that they were happy to drink surrogates like hair oil and methylated spirits. These factors contributed to the persistence of chronic chest infections, boils and infected wounds, whilst very high infant mortality was the result of ignorance and what we would consider as negligence and neglect.

More than a third of children changed families before the age of five. Most were probably unaffected by this, but they lost their inheritance rights, sometimes the adopted children acted as little more than unpaid servants, and sometimes they were doomed to a nomadic existence wandering between families. Young people were expected to take on family responsibilities at a very young age, which meant that a young couple with an extended family could realistically anticipate quite quickly relinquishing work and enjoying a life of relative ease. Most young people had their first sexual experience by the age of twelve or thirteen, and it seems they saw no reason to be careful of syphilis as they had all had this disease for a long time, in most cases from birth.

These issues of diet, health and family life, together with the existence of a large population of dogs living permanently on the verge of starvation because the islanders preferred to let them scavenge rather than take the trouble to feed them – these issues are perhaps no more or less than the social issues that in varied forms affect any small society. They are also, however, things easily unseen or disregarded by outsiders seeking evidence of an earthly paradise.

In his conclusion, Danielsson casts no judgement on whether life on a coral island is better or worse than "civilised" everyday life. Instead – and perhaps realising that a popular book about "children in paradise" is all it would need to destroy a precarious culture he values so much – he concludes only that, on Raroia, people lived "a pleasant open-air life among cheerful, friendly children of Nature", something that was only possible because the islands were completely closed to foreigners by government policy and by reason that there was no land for sale. And even this is qualified by his recognition that "if the gates of paradise were opened to all those who had no qualifications but that they liked the place, it would naturally cease to be paradise". This turns out to be particularly prescient in light of the French government not long after choosing to "like the place" (the Tuamotu Group) for its nuclear test programme.

No "gates" could protect Raroia's exceptionalism in perpetuity, that even a small remote group of islands with few reasons to attract outsiders would, if for no other reason, produce younger generations for whom the lure of Tahiti and beyond was too great. This meant that diet would continue to deteriorate, drink would ruin more people, and unscrupulous traders – in this respect he is particularly critical of the Chinese – would have better communications and easier ways of exploiting the islanders. Worst of all, with Raroia exposed to a wider world, the islanders would fall prey to diseases, even common ones like measles and influenza, of which they were ignorant and from which they had little immunity. Children would be exposed to "more profitable virtues" of acquisition and consumption that would challenge a hitherto simple way of life, and so old values would lose their meaning, just as the time-honoured spirit of Polynesian friendliness and magnanimity would fade.

In the twenty-first century, Raroia has the vestiges of its traditional activities, but income is mainly from tourism. Since 2006, an airport has received irregular visits from Air Tahiti. Danielsson himself submitted a doctoral dissertation a few years after his stay, an anthropological study of Raroia and neighbouring islands in the Pacific; and he and his wife, Marie-Theresa, went on to become highly active and respected critics of French policy towards indigenous cultures, and of the use France made of its latter-day empire in the Pacific for nuclear testing.

Despite the measured conclusions presented by anthropologists, the notion of discovering some vestiges of an earthly paradise continues to exercise the popular imagination. Pentecost Island – one of eighty-three Vanuatu islands – was the subject of a survey based on the Happy Planet Index (produced by the New Economics Foundation, a London think tank), and in 2007 was declared on BBC national news to be "the happiest country on earth". Based on life expectancy, expressions of general happiness, and the size of carbon footprint (the survey was partly funded by Friends of the Earth), the United Kingdom came 108th and the United States a lowly 150th. Meanwhile, in California, a computer delivered up Upolu in Western Samoa as its ultimate definition of paradise based on stereotypes such as balmy weather, sandy beaches and a healthy environment. [15]

Island Retreats

Throughout the Western world, islands have been sought after as retreats from mainland values, and as means of seeking redemption closer to God. The coasts of Britain and Ireland have an abundance of off-lying monastic islands: Iona was an important stepping-stone in the spread of early Christianity, as were the Aran Islands. Fifth-century monastic occupation is evident in the remains of wells, cells, chapels and beehive dwellings on Scottish and Irish western islands. In the seventh and eighth centuries, hermits in Canna, the small islands off Barra, Harris and the Uists, St Kilda, the Flannans, North Rona and the Shiants selected sites with water and fertile soil. But they were also far out into dangerous territory. Here would be experienced nature in the raw, not because nature was held in a pantheistic relationship with God, but because its horror set them at God's mercy. Accordingly, "the sea itself terrified them. It was the zone not of divine beauty but of destruction and chaos. Only God and the saints could control it. Others were at its mercy." [16]

Indeed, penitence often seemed to demand the close proximity of Heaven and Hell. As Samuel Johnson noted in his journal through the Scottish Hebrides in 1773, "the religion of the middle ages, is well known to have placed too much hope in lonely austerities. Voluntary solitude was the great act of propitiation, by which crimes were effaced, and the conscience was appeased; it is therefore not unlikely, that the oratories were often built in places where retirement was sure to have no disturbance." [17] Living in cells elemental in the extreme, freezing and half-starved and confronting what was often a ferocious western ocean, the monks on Skellig Michael, fourteen kilometres off the south-west coast of Ireland, "may have thought that their last battle against the demon host might be carried to a conclusion. Cut off from all the easy ways of the world with all the issues of life simplified … they offered themselves up in a white martyrdom of utter privation." For many, the "martyrdom" was complete when Vikings sacked the island. [18]

The term "utopia" derives from the Greek *outopos* – *ou* meaning "not" or "no", and *topos* meaning "place". Therefore, its meaning – or its *implied* meaning – is an idealised place that does not or cannot exist. However, the word most often refers to somewhere sought after by people wishing to realise a more ideal world. Utopias are places, often communities, underpinned by progressive or retrogressive ideas about human values and notions of society, religion and economy. [19]

Islands off the British Columbia coast of Canada have received a seemingly disproportionate interest from utopists. Danes at Cape Scott on Vancouver Island and Finns on Malcolm Island attempted to develop socialist-cooperative communities. And Brother Twelve ruled over a community on De Courcy and Valdes islands in the 1920s with a distinctly self-interested world view. [20] Stephen Guppy's short story "Ichthus" captures the spirit of some of the more eccentric and flamboyant enterprises on a myriad of islands in the strait between the Canadian mainland and city of Nanaimo where "one could hardly pull in at a sheltered cove or knock at the door of a moss-covered cabin without coming face to face with the last of the Romanoffs or a disciple of Aleister Crowley". Some of these eccentric outcasts from the mainland were just plain antisocial, some claimed obscure religions or to be bringers of good news about some new faith, and some were just conmen who came and went in a moment. [21]

The British travel writer Jonathan Raban, sailing the coast from Seattle to Juneau, is confirmed in his suspicion that "the idea of creating utopia in the wild is programmed into the far-western imagination, and the Northwest coast was littered with such projects, started by Hutterites, vegans, Indian spirit channelers, survivalists, Christian sects so fundamental that even fundamentalists thought them eccentric." Within a few years loneliness, the raincoast weather and the relentless battle against the rainforest took their toll. What started off as abandoned logging camps and canneries would revert to nature again, sprouting ruins within "ruins within ruins, as successive bands of hopefuls tried and failed to make a go of it." [22]

The life and death of utopias is a significant theme in literature and, just as the first ill-fated "New World" colony was sited on an island — Roanoke Island in present-day North Carolina in 1587 — so fictional utopias have often been located in remote and inaccessible places. Like the biological closed systems that characterise small remote islands, utopias appear unable to survive in proximity to alternative societies. Thomas More's Utopia (1516) was located between India and Brazil, Tommaso Campanella's City of the Sun (1602) was in Ceylon, Johann Valentinus Andreae's Christianopolis (1619) was in the "Ethiopian Sea", and Sir Francis Bacon's New Atlantis (1627) was in the South Pacific. Aldous

Huxley's *Island* (1962) provides a celebrated account of the inability of a utopic society to co-exist with a powerful and proximate continental neighbour. [23]

Hell on Earth

Utopia – island – dystopia: from the dawn of history Isles of the Blessed have existed in an uneasy relationship with Isles of the Damned. Just as Christian theology provided a rationale for seeking heaven on Earth, so in keeping with its dualistic philosophy, there was reason for believing that hell, too, would have its earthly counterpart. For medieval Christians, hell was specifically located at the core of the Earth, and speculation surrounded possible points of access. Stromboli (off Sicily), Hekla (off Iceland) and the "Isla del Infierno" (probably Tenerife in the Atlantic) were candidates for this dubious accolade. [24]

The Isle of Spirits, sometimes charted as the Isle of Demons, in the Gulf of St Lawrence, was believed to be the restless home of drowned voyagers, whose dreadful cries and fearsome reputation discouraged settlement or even landing by white people. Unlikely as this description may appear, one should beware of dismissing out of hand apparently preposterous island geography like this as the product of imaginative invention and sheer mendacity. Closer scrutiny can yield unexpected surprises. It is quite likely that exploration of this part of the north-west Atlantic would have coincided with the nesting season for pelagic bird species. Islands would be home to a teeming mass of birds that would have made a noise compatible with contemporary expectations of hell! So it was an unpleasant place, this Isle of Demons, perhaps what is now called Hospital Island (sometimes Harrington Harbour) off the coast of Quebec – where Marguerite de la Roque, a French noblewoman, was marooned in 1542 for the sin of infidelity.

"Satanaxio Island" occupied an area from the mid-Atlantic to the Labrador coast on charts from 1436, and in 1507 another "Isle of Demons" was located at the entrance to Hudson Strait. Both were erased from charts by 1645. [25]

Perhaps it is the propensity for optimism in human nature that has led to the Islands of the Damned being erased from charts more readily than Islands of the Blessed. Yet damnable islands continue to attract the melancholy imagination of writers who view them with foreboding, but also with a profound sense of disappointment. Here are fruitless places, their very existence defying reason, and mystification like this is deeply unattractive to the rational Western mind. So, in the prelude to passing the Island of Rockall Act in 1972, a Labour peer and former merchant seaman described the subject of the last land-grab of British imperial history as "no place more desolate, despairing and awful".

Rockall exists precariously, challenged relentlessly by a sea of destruction. And it is this apparent precariousness that seems to inspire a melancholy imagination. Thus George Hugh Banning was repulsed by the geologically immature landscape of Socorro, 600 kilometres west of the coast of Mexico. Its thorny vegetation precluded close inspection of the island, which from afar resembled a half-burned pile of cindery waste, with even its smouldering subdued by puddles of inky rain. [26] While Ascension, made up almost entirely of lava flows and cinder cones, and with forty-four distinct dormant volcanoes, so demoralised Simon Winchester that he described it as "'hell with the fire put out' ... the earth is in its raw state, unlovely and harsh, and grudging in its attitude to the life that clings to it". [27]

But he is willing to offer alternative locations for his vision of hell, not least in the Aleutian Islands, in the North Pacific, which confront the existential extremes of violent wind, extreme cold, enormous waves and bitterly cold currents at "one of the least congenial places on the planet ... immense fog banks of density unknown elsewhere in the world. And as if to add to the misery of this awful place, the countries that abound on to it seem to glare at each other over disputed islands ... where no one ever cares to be sent, and where morale is low and tempers run high.... [Attu Island] is an utterly wretched place – permanently cold, eternally foggy, treeless, windswept, without beauty, calm or even the serenity of Arctic solitude. It is, perhaps, one of the worst places in the world." [28]

"The worst place in the world"? Even further north, the Pacific squeezes through to the Arctic Ocean through the Bering Strait, eighty-two kilometres wide. In the middle of the strait, Big and Little Diomede are separated by four kilometres, a channel occupied by the Cold War's "Ice Curtain" and by the International Date Line. Big Diomede thus becomes "Tomorrow Island" to Little Diomede's "Yesterday Island". Of Big Diomede, a military base during World War II, the border was suddenly closed in 1948 and the indigenous people were forced off to mainland Russia to avoid contacts across the border with the United States. It is about thirty square kilometres, and is occupied by about 100 military personnel, who operate a weather station and act as border guards. A glance at a map might suggest that here is an international incident waiting to happen, and some observers claim that helicopters are very readily scrambled; but for the indigenous people of Little Diomede the nearest thing to a critical situation can be resolved by the Russians shouting warnings across when they think their neighbours are coming too close as they fish through the winter ice.

Little Diomede is about a quarter the size of its brother with which it shares certain weather characteristics: cold summers and extremely cold winters, high snowfall, and summers of fog and cloud when winds blow consistently from the north with gusts of 100 to 130 kilometres per hour not uncommon. That the sea is frozen for half the year provides a means of escape from a minimalist landscape – a more or less circular island with a small, flat top hazarded by uniformly very steep, rocky and snow-covered slopes that reach the ocean at short cliffs. It is almost devoid of vegetation.

The two islands and the strait between conjure an image of "Ultima Thule", existing beyond the reach of civilisation; and the tiny bestilted Ingalikmiut Eskimo village of not much more than a hundred souls clings to the base of this hillside, looking like it could tip into the sea at any moment. The community is served by a helicopter pad on a tiny, artificially created nose of ground that breaks the circularity of the island.

Originally a spring hunting campsite, the site of settlement activity for perhaps 3,000 years, it was a centre for whale, polar bear, walrus and seal hunting, bowhead fishing and crabbing. It had a long a tradition of trading with both continents. Even quite recently their skin boats would travel to mainland Alaska, especially Nome, to sell ivory carvings and skins for supplies.

John Muir travelled through in 1881, and found the people "eager to trade away everything they had", living in a village perched on a steep rocky slope, houses seemingly nothing more than stone heaps, and with canoes stowed out of reach of dogs that would otherwise eat them. For Muir, the Diomede settlements were "the dreariest towns I ever beheld – the tops of the islands in gloomy storm-clouds; snow to the water's edge, and blocks of rugged ice for a fringe; then the black water dashing against the ice, the gray sleaty sky, the screaming water birds, the howling wind, and the blue gathering sludge". [29]

Look at the settlement today and that grim dreariness persists. Since the mid-1950s onwards, Little Diomede has gradually become a permanent rather than a seasonal settlement. About thirty wooden cabins were constructed in the 1970s and '80s by the state government, but only more recently have they been connected by stairways and boardwalks. There is no piped water supply, only two buildings have a sewage system, and waste that cannot be incinerated floats out to sea at the spring melt. Thirty-five per cent of the inhabitants – and nearly half over the age of sixty-four – live below the poverty line.

The Little Diomede Council has voted for relocation to mainland Alaska if conditions of access and housing do not improve, but a report by the Environmental Protection Agency in 2006 warned that better access could have a negative impact on the local traditions and identity of the native community. Every photograph of village schoolchildren shows them smiling and laughing, blissfully unaware as they learn English and not their native tongue, Inupiat, that they may soon be leaving Yesterday Island forever.

This "dreariness" that subdued Muir's imagination on Little Diomede seems to cling to the writing of so many travellers when they encounter what nags them as the raw, inexplicable pointlessness of extreme island settings. Even Captain James Cook, whom we associate with a prosaic "Admiralty style" of writing, described South Georgia, where he made the first recorded landing during his second circumnavigation of 1772–75, as a "country doomed by nature never once to feel the warmth of the sun's rays, but to lie for ever buried under everlasting snow and ice". [30]

These islands on the Antarctic fringe seemed to inspire a particular sense of desolation in the minds of usually rational sea captains. James Douglas described Macquarie Island in the deep south of the southern oceans as "the most wretched place of involuntary and slavish exile that can possibly be conceived". [31] And Charles Barnard described the Falkland Islands as a hellish place of darkness and desolation where, should one escape being dashed to pieces on its shores, this was likely to be "only a prelude to a more lingering and awful death". [32] Captain Barnard was speaking with some authority, and his response was based on more than an *en passant* sense of repulsion. He had the misfortune to rescue some shipwrecked souls who returned the compliment by marooning him and a small group of sailors. His "crew" eventually stranded him on a small island with virtually nothing, and only returned to seek his forgiveness after they had marooned one of their own.

Our irresistible interest in hellish islands sits well with the contemporary culture of pessimism, something the media in general is strongly attached to. The popularity of books about paradise being lost, troubled, plundered or occupied by a serpent attests to this. *The Guinness Book of Records* thrives on a propensity for league tables and has declared Palm Island – off the coast of Queensland and which it called "Devil's Island" – to be the most violent place on earth outside a combat zone. On the other side of the world in 1993, Canadian national television broadcast an amateur video made at Davis Inlet, on remote Iluikoyak Island, which is about seventy square kilometres and just a kilometre off the coast of Newfoundland. Dubbed "the Island of the Damned" by the media, it presented a grim history to explain equally grim images of six Innu children sniffing gasoline and saying they wanted to die. [33]

In 2012, it was reported that Corsica, with its fractured independence movement, had France's oldest population, its worst education record, its biggest disparity between rich and poor. With only 0.5 per cent of the population of France but with twenty per cent of all its "revenge killings", it was considered to be (proportionately) the most murderous and criminal place in the European Union. It eclipsed two other islands with reputations for assassinations: Sicily and Sardinia. But by 2013, it was the French island of Guadeloupe that was being reported as the nation's most dangerous place to live, with more murders than Corsica and Marseilles. [34]

A dystopia is a repressive society in which art, religion, science and literature are distorted in the cause of a perverted view of social justice. Dystopian literature often presents a corrupted vision of utopia. Thus, Bacon's *New Atlantis* called for state-funded research and the nurturing of a scientific elite to enable the systematic subjection and control of nature. However, I use the term here in a more general sense, based on its etymological derivation, *dys* from the Greek "bad" or "difficult", and *topos* meaning "place". It concerns environmental experience that negates notions of the good life, and where life instead is bound inextricably with uncertainty and anxiety. [35]

We have visited Clipperton Island in Chapter Three in consideration of its fragile geography, but it is worth making it a port of call again because its settlement history conforms so well to this definition of dystopia. This speck of coconut palm and sand-encircled lagoon, 1,100 kilometres from the south-west coast of Mexico, is the only coral atoll in the entire Eastern Pacific. It is tempting to assume here would be a fine place for any voluntary maroon to live their island dream. But this assumption would be wrong.

Shaped approximately elliptical, and just six square kilometres, its stagnant freshwater lagoon is encircled by a low-lying strip of sand and coral rock, averaging about 150 metres in width, 400 metres at its maximum and narrowing to fifty metres in the north-east, where saltwater periodically splashes into the weed-covered lagoon. There are no fish in the lagoon, just millions of isopods that swimmers claim to have a nasty sting. Clipperton has a tropical climate, with a rainy season from May to October when it is subject to tropical storms and hurricanes. There is no natural harbour, and the surrounding reef dries at low tide and is often pounded by a Pacific swell.

A barren strip of rock then, in the middle of nowhere, but not quite, for colonising powers have always been diligent in deriving capital from the most unlikely places. Between 1849 and 1858, the Mexican, US and French governments laid claim to the atoll. By 1897, the American Guano

Company had been actively mining for a few years when Mexico decided it no longer desired Americans living so close to its coast and reasserted its claim to sovereignty by evicting them.

Undeterred, a British guano mining operation decided to try its hand in 1899, and by 1906 the British Pacific Island Company had acquired the rights to guano deposits. They built a mining settlement in conjunction with the Mexican government, which reaffirmed its claim to the territory by building a lighthouse in the same year. About 100 people occupied the island at this time, but by 1910 the British decided there was no future in the enterprise and withdrew all their employees, save for one caretaker.

In the meantime, Mexico garrisoned the island with thirteen soldiers, together with a governor, wives, servants, children and a lighthouse keeper. There were about twenty-six people in 1914 when a US ship was wrecked on the island. Rescue was effected quickly, and the ship's captain advised the Mexicans to leave too, an offer that the governor, Ramon Arnaud, declined. Instead he ordered the British caretaker and his family to leave with the Americans. With Britain's interest on Clipperton curtailed, and with a civil war increasingly consuming Mexico's attention, it seems that the little community was simply forgotten. The supply ship, once assured to arrive every two months, just stopped coming.

The story from now is harrowing in the extreme, and bears marks of authenticity by common threads that are present across a number of sources. It is possible that the marooned-cum-stranded community could have survived on a diet of fish, birds, bird eggs and a few coconuts, but it seems that the vegetable garden tended by the British had been allowed to revert to nature, which meant that the meagre diet was vitamin deficient, and it was not long before the adult men in particular began suffering from scurvy. Suffering a terrible death, and then buried deep in the sand to avoid being eaten by crabs, the last of the men – bar one – perished on the reef when their boat was overwhelmed whilst seeking to alert a passing ship. Just hours later, Governor Arnaud's widow went into labour, having to take refuge in the tiny basement of their house whilst a hurricane rampaged through the tiny settlement.

Things could hardly get worse, but they did. In the aftermath of the storm, and as the women and children picked through their damaged houses, the thus-far anonymous figure of Victoriano Alvarez – the lighthouse keeper – enters the tragedy. His reclusive ways may have enabled him to out-survive the rest of the men; his African origins may have meant that he had suffered discrimination from British, Americans and Mexicans alike; he was probably mentally ill for, so it is said, lighthouse-keeping was notorious for causing madness.

Destroying all of the islanders' weapons, except his own, Alvarez – now the self-styled king of the island – embarked on a reign of terror, rape and murder. Things staggered on for nearly two years, the women and children sick, malnourished and in dreadful fear until they found a way to strike him down with a hammer and slash at him repeatedly with a knife. The day was July 18, 1917.

One really needs to take a deep breath now, but there is no time. Even as the blood was drying, the USS *Yorktown* hove into view and succeeded, only at the second attempt, in getting a boat ashore. Three women and eight children were ferried to the warship, which broke off searching for German U-boats and set a course for Salina Cruz, Mexico. Alvarez was left to the crabs.

In deference to the future lives of the women and children, the official report of the rescue divulged no information about the death of the lighthouse keeper. For many years, the navigator lieutenant and the captain kept their silence, whilst the story of the eleven survivors passed by word-of-mouth to become well-known along the Pacific coast of Mexico. [36]

Clipperton is a fragment of land in the immensity of the ocean. Four nations, various guano companies, military personnel and their families, and a lighthouse keeper all play their part in its tragic story, but only the three women see the tale out from beginning to end. If there is an even more down-to-the-wire dystopia, it is *one person* alone on a rock, a person whose very existence is in peril at every moment. In William Golding's novella *Pincher Martin*, an apparently doomed British sailor is cast violently upon a rock somewhere in the mid-Atlantic. For Golding, this anonymous, most terrifying speck of land is a stage set for an allegory of purgatory where the bitter taste of memory after a selfish life culminates in alienation, loss of identity and loss of sanity. Like Defoe's Crusoe and almost all of his imitators and re-interpreters, Martin busies himself as best he can, trying to keep the demons of terror at bay. But unlike Robinson Crusoe, no amount of toil is going to save Pincher Martin from damnation; and in the final twist to the story, where it appears that his rock-bound terror is a kind of pre-death vision, he is found cast up on a beach, drowned even before he could kick his sea boots off. [37]

A dystopia, then, is an environment where events seem devoid of meaning or where meaning is expressed only in fear and anguish. Pincher Martin's rock is a terrifying place and his experience is likewise. But physical and experiential environments do not always coexist in a rational, cause-and-effect sort of way. Thus, for example, a group of political prisoners leaving the penitentiary of Robben Island for the last time see it as "a green and picturesque land in the ocean, the harsh monotony of its internal life totally hidden by its outer physical beauty". And, they go on, "may all who live on you be liberated, and may you go to hell". [38]

Likewise, Norfolk Island, which also served as a penal colony, has been described as having a *delusive* beauty, "an apparition, a rolling cap of green meadow and spiring trees, raised out of the Pacific on pipes and pillars of basalt as though offered to one infinite blueness by another. ... One sees nothing but elements: air, water, rock and the patterns wrought by their immense friction. The mornings are by Turner; the evenings by Caspar David Friedrich, calm and beneficent, the light sifting down towards the solemn horizon." [39]

It was this "delusive beauty" that puzzled Supreme Court Judge William Westbroke Burton, who held court there in 1834, for here on Norfolk was a radical discrepancy between the island's physical and experiential environment. And indeed he was clearly distressed that the Romantic belief held by so many educated men at the time – that landscape had a therapeutic effect on human character – had to be suspended on Norfolk. Missionaries, too, pondered the inverse relationship between the Godly nature of the landscape and the Devilish hopelessness of the prisoners, no doubt gaining some solace in the belief that here was a modern representation of Sodom and Gomorrah, where God's wrath was visited upon those guilty of the foulest crimes.

So high court judges and missionaries, men otherwise claiming an educated, charitable and sympathetic view of the travails of human kind, felt at ease in ignoring the fact that many of the prisoners were victims of a brutal system, designed to stamp out all hope of redemption and for whom the beauty of the physical environment was a great superficiality. [40]

A history of Robben Island prison, with its paupers, chronically sick, lepers and convicts of all denominations, is subtitled "out of reach, out of mind", an epithet that suggests that by making something physically remote we can isolate our minds from it. [41] In Europe during the Middle Ages, leprosy was associated with moral degeneracy and a fear that the victims risked contaminating society at large. Leprosariums were established on islands more as isolation units than treatment centres. Many varieties of the disease are now recognised as not highly contagious, but even in the twentieth century isolating patients continued to be perceived as an appropriate "solution". Such a response is illustrated in the case of Little D'Arcy Island – eight kilometres from Victoria, British Columbia – which was the site of a leprosarium between ca. 1890 and ca. 1926. The scant records for this period suggest the island was more of a permanent quarantine station than a hospital, where there were very few efforts made to give medical treatment or to relieve pain. That many of the lepers were of Chinese origin is almost certainly a factor in this gross neglect. [42]

Marilyn Bowering's research for her excellent book *To All Appearances a Lady* revealed that food was delivered every three months, and that there were no medical supplies except opium. Locating a leprosarium on a small island, she concluded, enabled those responsible to conveniently ignore the existence of the occupants, and to act as if they were "disappeared persons". Doctors seldom landed, but occasionally shouted instructions from a boat. In the late nineteenth century, people in the adjacent community of Cordova Bay noticed a fire on the island, but nothing was done to investigate the problem for three days. The raising of a flag was a sign indicating that assistance was needed, but on one occasion when somebody was seriously ill, six weeks elapsed before there was a response, by which time the person was dead. Bowering postulates that "islandness" presents people on the mainland, no matter how close, with the option of constructing unconsciously an ambiguous and skewed perception of land offshore. This process of "blanking out" is evident here in a perception of lepers who are not seen, *ergo* they do not exist. The mental dexterity to surround islands in an "inpenetrable mist of our imagination" is expressed well in a passage from her novel:

"D'Arcy Island, the leper's island. Where there was no traffic. No boats landing or casting off. No passengers, no observers or witnesses. ... It was not a subject much talked of: it tended to be bad for morale. And the lepers were all Chinese, so far as he knew. Once sent to the island they ceased to exist for the rest of the world. Even passing ships avoided their shores, arcing away from the coastline where the lepers had their colony. People were superstitious about it, as well as frightened. The Indians said the island was cursed." [43]

Islands of Isolation

"Penitent" and "penitentiary" share etymological elements, but they perform contrary roles in island society. A penitent is one who acknowledges their sinfulness and volunteers to remove themselves from the greater society. They choose to seek God's forgiveness in a remote location where earthly temptations are few. This is not for the faint-hearted, and for those less inclined to make this sacrifice, society has invented the penitentiary as a means of coercing the sinful – usually quite unsuccessfully – into penitence and out of the way of temptation. [44]

So isolation has been sought by some and imposed on others; some have sought a new world order while others have been excluded from the existing order lest they damage it. This need to isolate the unwanted and the unworthy has been met by using islands for imprisonment, exile, marooning, the isolation of medical conditions like leprosy, plague and cholera, and for the inspection of would-be immigrants and asylum seekers. The number of islands that have fulfilled these roles is too numerous to detail, so we will dip into some specific examples to exemplify the issue.

In Rome, Suetonius, in *The Twelve Caesars*, records Augustus and Livia condemning their daughter Julia and grandson Agrippa Postumus to very small "prison islands" in 6 or 7 AD. This was Pianosa, just ten square kilometres, close to Elba off the Italian coast of Tuscany. (In fact, Pianosa has been used as a prison for a period spanning some 2,000 years. Between 1958 and 1998, it was a maximum security prison, during which the Italian government announced, after the murder of a leading Mafia prosecutor, that the island would have a special facility to hold especially dangerous Mafia criminals.)

The more-favoured Tiberius was exiled to the much larger Rhodes for seven years. History may judge that the world would have been a better place if his sentence had been extended indefinitely. However, having direct experience of this nasty form of punishment, Tiberius was well placed to choose exiling islands located between Rhodes and Naples to banish Agripinna and her two sons to Pandataria (now Ventotene), and to starve Nero to death on Pontia (now Ponza). Other Roman officials were exiled to Corsica and Sardinia, and to Amorgos and Seriphos in the Cyclades Group. This tradition was continued in Italy where, at the beginning of the twentieth century there were prisons in the Ponza Islands, and Aeolian Islands, and on Procida, Elba, Pantelleria, Ustica and Lampedusa. [45]

The most systematic use of "foreign shores" as prisons was in the establishment of convict colonies by Britain in Australia. The initial aims included the reduction of England's "criminal class" who might be reformed in the process of colonising new lands; but by the 1830s the emphasis was on deterrence, whereby the guilty would suffer a fearful fate, the news of which would terrify the innocent away from crime. On Norfolk Island, all pretence at reform was dropped, and commandants were invested with arbitrary powers enabling them to order hellish punishments. This was a prison of such ferocity that, according to the Reverend Sydney Smith of the *Edinburgh Review*, "men recoil in horror – a place of real suffering painful to the memory, terrible to the imagination ... a place of sorrow and wailing, which should be entered with horror". [46]

The sailing guide for waters around the Italian island of Ustica describes its history as "littered with accounts of massacres and mayhem". The Greeks called the island Osteodes, the Island of Bones, on account of 6,000 mutinous soldiers from Carthage who were abandoned on the island to die of hunger and thirst. In the eighteenth century, the Bourbons attempted to colonise the island, only to have the colony massacred by pirates, with only two escaping to tell the awful tale. [47]

Lampedusa has a long and continuing history as a prison and "holding centre" for asylum seekers. Located as it is, little more than 100 kilometres from Tunisia and closer to Africa than to Italy, as recently as September 2011, hundreds of immigrants clashed violently with police and residents. So many migrants were landed on the island in 2011, following political upheaval in Tunisia and Libya, that the local population of 5,000 was heavily outnumbered by the migrants. The tragedy of African migrants being prepared to sacrifice their lives to reach the more migrant-sympathetic nations of northern Europe became international news in October 2013 when 366 people died within sight of Lampedusa after their boat caught fire and sank. Between June and September 2014, the United Nations High Commissioner for Refugees estimates that 2,200 people died – 750 in the space of a week in mid-September – and that a total of 3,500 perished in 2014. In just a few days during February

2015, 2,700 migrants were rescued from the Mediterranean, and in the same period at least 300 drowned. Mostly from Eritrea, Somalia and Ghana – and increasingly from Syria, many take ship in Libya and Tunisia – and become part of the huge influx of migrants who have struggled towards a new life in Western Europe in the last five years.

In October 2008, the Greek island of Agathonisi, with a population of just 150, was trying to cope with an influx of 4,000 migrants. These included people from as far away as Afghanistan and Iraq, who were dropped on the island by smugglers from nearby Turkey. Some were moved on to the island of Patmos, which then closed its docks to any more, as migrant numbers overwhelmed island inhabitants. Humanitarian concerns notwithstanding, the island's tourist representatives were clearly dismayed: "We are totally against the transformation of sacred Patmos into a ghetto for hungry immigrants from Africa and Asia," wrote the island's union of hoteliers to the prime minister. "Our island cannot be promoted as a destination for high-end tourism on the one hand, and on the other allow hundreds of illegal immigrants to wander around hungry and dirty." [48]

Juxtapose this view from the streets of "sacred Patmos" with that of Pope Francis, who chose Lampedusa as the venue for his first trip as pontiff. He addressed the European Parliament in Strasbourg in November 2014 and expressed his solidarity with poor migrants and workers in stating that "we cannot allow the Mediterranean to become a vast cemetery.... The boats landing daily on the shores of Europe are filled with people who need acceptance and assistance." As of 2017 there seems little evidence that the migrant crisis in the waters between Africa and Europe is relenting.

Greece has recognised that migrants "need acceptance", but there has been violence, too, as on Lesbos and nearby Chino. Here, as on other islands, people in desperate need have made a perilous journey and are still faced with an uncertain future, and they are held in close proximity to an island population that feels its security and future too is imperilled by the threat of living in reduced circumstances.

The island of Spinalonga, off the north coast of Crete, was Greece's main leper colony from 1903 until 1957. The Greek island of Leros has, for centuries, been used for outcasts – lepers, convicts and political prisoners, some incarcerated for as long as twenty-five years. Like the Canadian island of Iluikoyak, it has also been described in the media as an "island of the damned". A 1984 European Commission team reported the plight of mental patients on the island of Leros as "an accumulation of human misery" on an isolated "human dumping ground for the unwanted". [49] A Dutch psychiatrist described patients "dumped in this isolated place to rot. You might have killed them then. But most awful is the fact that that they are simply waiting until death comes. This is the destruction of the spirit, of the man not the body." [50] In 1992, it was reported that, despite £50 million being provided by the European Union, there were still young mental patients chained to their beds, left unattended for long periods, and with their human rights violated on a daily basis. [51]

Chapter Eight

An Island Mentality

The notion that geography determines the nature of culture, society and behaviour has long fallen out of favour. Geography influences how we respond to the world about us, but so too does a range of factors from the cataclysmic to the serendipitous. So we will need to tread gently in exploring the extent to which island geography influences the attitudes of those who appear to be beguiled by them and the behaviour of people who live on them.

Social psychology, and indeed common sense observation, teaches us that making generalisations about attitudes and behaviour is perilous. It seems that the collective of island experience is beset with paradox, the coexistence of apparently contradictory elements so that island existence often seems to dangle uneasily between notions of heaven and hell, paradise and purgatory, utopia and dystopia, and topophilia and topophobia.

But troublesome issues like this have in no way deterred writers from attributing unique characteristics to the experiential landscape of islands.

Like Nowhere Else

For H.E. Bates, men long to possess islands as they long to possess lovers, but in both they are inevitably disappointed, not only because there are never enough islands to satisfy man's great island appetite, but because expectation is so easily deceived. Islands may be so easily perceived as beautiful, but it is a tempting, beckoning beauty that can imprison those so beguiled. [1]

This deceptive quality of small islands and the disappointment of their possession is about the enisling sea, the very thing that defines the island and which can be profoundly unwelcoming. The sea ensures it is difficult to feel really secure as monarch of all one surveys – indeed, the island is quite likely to reverse conventional roles of ownership and take possession of the islander. [2]

Adam Nicolson, whose family has owned the Shiant Islands in the Scottish Hebrides and who could lay claim to being the monarch of all he surveys, recognises this in describing his islands, which can be as gentle as they can be brutal, that can be like fickle lovers seducing and spurning on a whim. They are a place where "life was so thick, experience so immediate ... [and] the barriers between self and the world so tissue-thin". And it is the sea that "elevates these few acres into something they would never be if hidden in the mass of the mainland. ... They are defined by it, both wedded to it and implacably set against it." [3]

By definition, a small island has a high proportion of coastline to landmass, with a hinterland where the influence of the sea is all-embracing. This sea is where human evolution began, but as land-dwellers we evolved by casting off our marine dependence. Whilst it remains a great provider for life, through evolution we have escaped from it. In terms of survival, it is now perceived as alien territory, and we are attracted instinctively to a different habitat – land, an island set in an otherwise hostile environment. But beyond the basic need to survive, there is perhaps a remnantal affinity with the sea, a yearning to be close to it just as we once needed to stay close to our mother, with the sea reminding us of that amniotic fluid in which we were once enwombed.

The ocean, then, defines the island just as it challenges its existence, and the island provides shelter from the storm, but shelter that is forever being threatened by the very thing that gives it existential reality. In this way, as Nicolson again emphasises, an island is "life set against death, a life defined by the death that surrounds it". [4] And so islands, like mountains, feed an appetite for the absolute: they can leave you feeling naked, shivering before the wind and wishing that life would end, just as they can make you feel more alive, nurtured and enriched by being there.

Certainly, many writers feel that the sea feeds their creative instinct, with the ocean having a powerful influence on the collective unconscious, something that, according to Carl Jung, is invested with mystery and inspires creativity. It was in this context that the writer W.D. Valgardson told me that the attraction of islands is that living close to the sea promotes an enhanced level of consciousness, a more fertile imagination, and a more creative and productive output. Likewise, the poet and novelist Marilyn Bowering describes the surrounding sea as a transparent boundary between this world and other worlds, a boundary that separates us from a more creative domain and which may be crossed when the mind is properly focused. Being on a small island helps this process of "focusing".

This supports the notion that there is an unconscious association between the island of geographical reality and the island of the inner self. Johann Wolfgang von Goethe, for example, with characteristic stridency, claimed, "he who has never seen himself surrounded on all sides by the sea can never possess an idea of the world, and of his relation to it". [5] Similarly, the central character in Knut Hamsun's *The Wanderer* expresses the consciousness-raising aspect of island-living when he states, "one day, I suppose, I shall weary of staying unconscious any longer; then I shall make my way once more to an island". [6] Just as a figure in a Dermot Somers short story proclaims, "I left this barbarianism, and fled to the west coast of an island off the western shore of an island way off the west coast of our island, where the cliffs prop the last arch of European sky above the rim of the western sea. I would achieve something colossal there, reeking of belief and transcendence." [7]

John Fowles, supported as he is by his islomane instincts, describes the irresistible siren call of islands, remote and empty where, if one does not discover Crusoe one may discover something about oneself. Islands, he claims, "haunt and form the personal as well as the public imagination" by provoking "a vague but immediate sense of identity. ... It is the boundedness of the smaller island, encompassable in a glance, walkable in a day, that relates to the human body closer than any other geographical conformation of land. ...Then there is the enisling sea, our evolutionary amniotic fluid, the element in which we too were once enwombed. ...There is the marked individuality of islands, which we should like to think corresponds with our own; their obstinate separateness of character, even when they lie in archipelagos." [8]

"A Particular State of Mind"

In terms of how island landscapes shape their inhabitants, both socially and psychologically, Fowles suggests a paradox: "On the credit side there is the fierce independence, the toughness of spirit, the patience and courage, the ability to cope and make do; [and] on the debit side the dourness, the incest, the backwardness, the suspicion of non-islanders ... all that we mean by insularity." [9]

The Island

Whilst islanders might see themselves as territorial, self-reliant, and keeping mainland values in clear view but at arm's length, outsiders could view this particular state of mind as insular, subversive and even duplicitous. [10]

The qualities that make up this "particular state of mind", and the paradoxes inherent within it, are expressed too in David Guterson's novel *Snow Falling on Cedars*, set on an island in Washington State. The narrator is a mainlander but is reluctant to be too critical of islanders, recognising as he does the limits of a world always surrounded by water. Here there was "no blending into anonymous background, no neighbouring society to shift toward" so islanders were obliged to tread gently as "no one trod easily upon the emotions of another where the sea licked everywhere against an endless shoreline". These muzzled emotions might enable islanders to live in apparent equanimity, but the price to be paid was that too much was held in, unsaid, as if they were holding their breath in "regret and silent brooding. ... Considered and considerate, formal at every turn, they were shut out and shut off from the deep interplay of their minds. They could not speak freely because they were cornered: everywhere they turned there was water and more water, a limitless expanse of it in which to drown." [11]

Guterson's islanders, then, existing in intimate proximation with their neighbours with "regret and silent brooding", could easily be considered obstinate, and tenacious to the extent that they will cling on to an existence off-islanders cannot apprehend.

The realities of having "no neighbouring society to shift toward" are evident, too, 9,000 kilometres south of Guterson's islanders, on Pitcairn Island, where Dea Birkett, in

Serpent in Paradise, reflects on the islander Alison, for whom the uniqueness of her island life makes her both very happy and profoundly depressed. This is a "uniqueness" that was "*predicated on its isolation.* You couldn't move on to the next town, try another resort up the coast road. This was it – all you saw, all you did, everyone you met was all you were going to see, do and meet, until you left the island, probably for ever." (The emphasis is mine.) [12]

This vision of a kind of point-blank reality, with its undertone of melancholy, seems strangely neglected by those seeking the meaning of island life. Instead, a more optimistic view prevails. The ever-enisling sea is a constant reminder that life is constrained, separate and distinct from the hubbub of the mainland, where much can be taken for granted. This means that small acts, which meet immediate needs, take on significance. Protected from the critical eye of the mainland, small island life can beguile one into believing one can simplify aspirations and find a role that is satisfying in its purposefulness.

"Encompassable in a glance", and remote from sources of mainland influence – here is a "person-sized landscape", a place for meaningful relationships uninterrupted by the complications of the mainland. As at the hub of a wheel with its spokes pointing out in all directions, the islander's view is expansive in circumference. Here is a place where the physical and social landscape may promote intimacy and solidarity, and throw into relief the isolation of mainland life with its mass of humanity enmeshed in complicated remote social systems. [13]

It is the anticipation of discovering this comfort of intimacy that can make tolerable the initially daunting prospect of arriving at an anonymous speck in the enormity of the ocean. For Dea Birkitt arriving at Pitcairn, living on a "speck" of just five square kilometres meant, in her mind's eye, that she would quickly be knowledgeable and confident with the island's geography and soon make a home amidst the tiny population. She never did succeed in becoming "a member of the Pitcairn family"; but she did develop an intimate relationship with the island's geography, recognising as she did the manner in which the islanders made sense of location by naming everything after themselves and so giving every curve in the road, every rock pool and every tree a distinct identity. [14]

It is worth spending more time on Pitcairn because of its particular history of human occupation. This group of islands is about 6,600 kilometres west of Panama, and 2,150 kilometres east-south-east of Tahiti. Although there were a few remnants of previous occupation when mutineers from HMS *Bounty* landed there, our knowledge of the island dates from that day in 1789, when nine mutineers together with six Tahitian men, twelve women and one child arrived. Fletcher Christian was their nominal leader. What they knew was that Pitcairn was difficult to land on and that it was far from established shipping routes. What they probably guessed was that Pitcairn was far enough away from archipelagos where the seafaring Polynesians were likely to find them; and what they did not know was that although the island appeared on charts, it was inaccurately placed so it appeared about 200 nautical miles from its actual location.

Accounts vary, but there is consensus that relations were not good from the outset. The mutineers had first made landfall in Polynesia, compared with which Pitcairn was a less attractive proposition; and it is likely that the indigenous

people felt betrayed over promises of a new life on a bountiful island. It is likely too that the mutineers were edgy and
anxious about whether any haven could be secure from the
prying eyes of the British Navy, which would be seeking its
own form of retributive justice.

A particularly brutal internecine war seems to have
occupied the first few years, so that little more than ten years
after HMS *Bounty* was burnt offshore to destroy evidence of
their existence, only John Adams and the women were alive.
It is tempting to hazard that the Europeans were a rabble
who, free from the controlling influences of naval discipline,
reverted to base instincts. One can say with some degree of
confidence that, once the mutineers had got rid of Captain
William Bligh, their behaviour was more an imitation than
a departure from the perceived savagery of his rule. This
estimation reflects the conclusions of the Pacific historian
Gavan Daws, who is more strident in his assertion that the
mutineers' idea of happiness "was to control others without
seeking controls on themselves. In these civilised men there
was savagery. They kidnapped Tahitian men and women
and took them to Pitcairn Island. They forced the men
to do hard labour, and quarrelled with them over women.
The human bill for all this excess inevitably came due for
payment. Liquor was being distilled and drunk on Pitcairn,
and a great rage brewed up over race and sex and domi-
nance. It led to plots and ambushes and axe murders, and a
welter of blood and destruction." [15]

Out of this turmoil emerged something that, on the
surface, was quite remarkable. In 1808, little less than twenty
years after the mutineers landed, the first vessel – under
American whaler Captain Mayhew Folger – discovered

the settlement. He described a population of thirty-five "very humane and hospitable people". In 1814, Captain Philip Pipon of HMS *Tagus* described forty-five people in good health and apparently happy under Adams' rule. And Captain Frederick Beechey of HMS *Blossom*, the third ship to touch at Pitcairn, in 1825, was equally impressed with sixty-six people living in harmony and happiness, their "moral, religious and virtuous conduct ... ascribed to the exemplary conduct and instruction of old John Adams". [16]

Sentiments like these were repeated numerously in early descriptions of the islanders' conduct and community, and this undoubtedly helped John Adams who, during these early visits, spun tales of the mutiny in which he was a coerced participant. Uncharacteristically for the British Admiralty – which usually hunted down mutineers to the ends of the Earth – Adams' story was either accepted or ignored. And on Pitcairn – where he was patriarch to a substantial family calling him "Father", and where he was "Commander-in-Chief of Pitcairn's Island" – his story was unlikely to be challenged.

But it is very possible in light of later events that the years 1800 to 1829 were not as harmonious as most observers were led to believe. Young men grew up under the powerful influence of Adams, but they could not escape the recent history that had riven family against family in bloody feuds. So when Adams died, they seem never to have been able to agree over a successor from one of their own men, and fell prey to the machinations of a number of adventurers and imposters, outsiders who divided and ruled. Adams' death heralded a period of anarchy in which, according to a whaling surgeon, "drunkenness and disease were amongst them – their morals had sunk to a low ebb – and vices of a very deep dye hinted at in their mutual recriminations". [17]

In the ensuing years, Pitcairn always had a dominant male figure, supported by a power base of influential men who controlled every aspect of life; and if their power was ever in doubt, it could be restored in the knowledge that the longboat – the Pitcairner's vital link to a larger world - could not be launched without them.

In 1852, the Reverend W.H. Holman spent nine months on the island, the first outsider to stay longer than a few days. Claiming that the island was suffering extreme soil erosion, that the climate had become too unreliable to sustain a population, and that the islanders had become indolent and obstinate too, he concluded his report to the Admiralty by stating, "there can be no doubt that the golden days of the Island and Inhabitants of Pitcairn are already past and gone. Everything I saw convinces me that they have been gradually retrograding since the time of John Adams." [18] Holman recommended their immediate evacuation to another island.

In 1856, the Pitcairners were resettled on Norfolk Island, 5,500 kilometres to the west, and it is testimony to the pull of their original island home that, within three years, sixteen had returned, followed by another twenty-seven in 1864. Few ships visited the island during this period; it took time for the resettlement to filter through to the maritime world, and the whaling industry had moved away from the area. These were difficult times for the Pitcairners, and in 1872 family and friends on Norfolk Island even tried to persuade them to evacuate again, the only condition being that all had to come. It was only because a few refused that Pitcairn clung on to its population.

It was only with the completion of the Panama Canal in 1914 that Pitcairn's location became something of an advantage, being almost exactly equidistant in the vast new open passage between Panama and New Zealand. Large passenger ships began to call on the island once a week until well into the 1960s, when long-distance air travel began to be important. It appeared that the population of Pitcairn may have been dwindling slowly but, tenacious as ever, they were finding new ways to survive.

In the mid-1980s, the British government spent £2 million on harbour construction. But more often, development was characterised by struggles with officials far away in London who were often perceived by islanders and visitors alike to be mean and negligent. The lack of social services was evident in quite extraordinary ways. Simon Winchester recounts a surgical operation that took place, probably in the early 1980s, performed by an untrained pastor, on a seriously ill patient, using hand-forged, home-made instruments, and guided by a surgeon speaking by radio 13,000 kilometres away in California. For this to be necessary, one could conclude that Pitcairners might feel ill cared-for by Whitehall.

In the late 1980s, a retired, eccentric coal mining engineer named Smiley Ratcliffe, from Frog Level, Virginia, offered to buy Henderson Island, which is part of Pitcairn's four-island colony. He wanted to set up a "centre for Pacific eugenics" and, for a ninety-nine-year lease, he offered the British government $5 million, a small airstrip on Henderson and a ferry that would allow Pitcairners relatively easy access to the outside world. Such a generous offer seems barely credible, even in the pursuit of Pacific eugenics but, taking advice from the World Wildlife Fund, the British government turned down the offer in favour of Henderson's population of apparently uniquely rare flightless rails, fruit-eating pigeons and snails. It is easy to imagine that policies like this would sit uneasily with long-suffering inhabitants of Pitcairn. [19]

It is likely, too, that Pitcairners felt some resentment over Norfolk Island, which had flourished after it was settled by families who had elected to remain after the evacuation of Pitcairn in 1856. Indeed, yet again, the Norfolk Island Council invited their distant relatives to resettle with them in 1964, fearing that the human resources of Pitcairn were insufficient to secure its future. They were concerned, for example, that the island would soon be without able-bodied men and boys to crew the long-boats, offload supply ships, trade on the ships and maintain the island infrastructure. By the time of Dea Birkett's visit in the mid-1990s, she counted only twelve men, including teenagers and grandfathers, doing this work.

Life on Pitcairn was never effortless. Its geography always sat uneasily with a stereotypical image of Polynesian paradise. The coastline is almost uniformly steep and rocky. Summers can be extremely hot and humid, sometimes making growing even vegetables mostly for local consumption difficult in near-drought conditions. In the rainy season, the few dirt roads turn to treacherously sticky mud. Until recently there have been no newspapers, televisions, banks, cars, cafés or bars. This might provide a basis for sustaining some notion of utopia, but an all-pervading Seventh Day Adventism proscribed alcohol, tobacco, various foods, and frowned upon a Polynesian culture with its subsistence ease, its cheerful world outlook, its music, dance and bright clothing. The gene pool is small, most people are related, and the population level is precariously low, which makes it difficult to fulfil the basic requirements necessary to sustain the community. Despite this, until quite recently even visiting the island has been notoriously difficult, with council meetings almost always rejecting even the most well-informed and innocently motivated applications. In so doing, the islanders appeared to resent anything that could be construed as outside interference and to focus upon themselves, even as they nurtured feelings of being neglected by their British administrators.

There was a courthouse and a jail used as a convenient store for miscellaneous articles. And since crime was never known, there was much comfort to be gained in the knowledge that, in a world of change, there was always the "constant rock of Pitcairn". [20] But a hundred years earlier, Amelia Young, a descendant of the original settlers, had unknowingly offered a prescient warning that human nature is as constant as the "rock of Pitcairn, but not always pleasant to behold". She observed it would be a mistake to assume that, "because so remote from the rest of the world, no vice or sin of any kind mars the character or degrades the reputation of those who dwell so secluded from the world". [21] Indeed, her words take on profoundly revelatory significance when a series of appalling events came to light, events that focused world attention on Pitcairn like never before.

The early years of Pitcairn were marked by brutality and fights over Polynesian women who had been abducted from Tahiti. Testimony presented at the trial of seven men on the island and six abroad suggests the abuse of young girls took place over three generations, and anecdotal evidence suggests that it had been commonplace amongst the great-grandparents of families still living on the island. And it is highly unlikely that the crimes, even if not perceived as such in the island culture, were hidden. Dea Birkett, who visited the island more than ten years before the child abuse was revealed to the outside world, noted that nothing was anonymous on Pitcairn because islanders lived in such intimate proximity with each other that everything was visible and, as in a family, small acts could take on a huge significance. [22]

The abuse may have continued for much longer but for a British police officer – the first to be temporarily posted to Pitcairn in 1999 – beginning an enquiry into a complaint. Eventually a far-flung investigation led to police interviewing women who had grown up on the island and who were living in New Zealand, Australia, Norfolk Island and Britain. Some of the accused men had, in the past, been invited by the Seventh-Day Adventist Church to the United States to share in the centenary celebrations of Pitcairn's embrace of the Adventist faith in 1886. Here they were toured as an exhibit from a South Seas island billed as "the perfect Adventist community, free from earthly sins and waiting for the second coming". [23] Even after he was charged, and while still enjoying anonymity, Steve Christian – the island's chief representative in travelling abroad to promote the island – went to New York and addressed the United Nations committee on decolonisation.

Six men were convicted and were given very lenient sentences tailored to the needs of the island, as the

Chief Justice in the case explained. They began serving sentences in a new prison in late 2006, and by 2010 all had served their sentences or had been granted home detention status. "Home detention" means that sex offenders and paedophiles were not treated as outcasts, condemned to banishment from Pitcairn, or even isolated from the young people they had abused. Indeed, with few exceptions there was no sense of having done something wrong, and for a significant group on the island related to or close to the offenders, there remained a powerful resentment that the trial had been a miscarriage of justice. And they were not alone. The myth of Pitcairn – this tiny fragment of land lost in the immensity of the ocean and with its colourful history of a tiny group of souls clinging on to life against all odds – meant they received support from Pitcairn aficionados from California to London to New Zealand.

For many people, in places where convicted paedophiles would not be welcome, Pitcairn was perceived to be *different*. For them, the islanders were like other isolated indigenous groups that practised culturally accepted rights of sexual initiation and rights of passage, "ceremonies" that are only considered unacceptable within the more "universalised" code of moral behaviour held by far-away modern societies. For people who believed this, here was a brave community, surviving through self-reliance and solidarity, which was being broken apart with its young men persecuted by a heavy-handed British administration. And this administration was only rousing itself because generations of child sex abuse was finally there for the world to see.

It has been observed that small islands can exert a

transcendental influence on human nature. But on Pitcairn there appears scant evidence to support this view. Rather, the original settlers arrived as mutineers, adventurers, tough men of dubious moral character bolstered by strong prejudices and beliefs. The trials for rape and sexual assault of children reveal a process of learned behaviour generation through generation. No matter how much they may have detested one another, the existence of the community depended on the men, and so did the women and children. If the majority of men were so inclined, children could be almost routinely assaulted and raped. The minority objectors would risk being marginalised and one step closer to helplessness.

This crude male power dynamic could be reinforced on Pitcairn because it is one of the remotest inhabited places on Earth, which until recently has not even had a scheduled boat service. So the elected male elite could operate a closed system with visitor requests usually turned down, or accepted for only short stays. And with visitors being welcomed by what they might easily perceive as exotic, charismatic machismo, it was easy to sustain the myth of Pitcairn, especially under the droopy eye of a British authority that seemed happy to let sleeping dogs lie.

Does this mean then that there is darkness in every one of us and that without the trappings of civilisation, we fall prey to brutish instincts? The history of human occupation on Pitcairn Island culminates in revelations of generations of brutish child abuse, but it does not answer this question, a question that has occupied social philosophers and psychologists for centuries and which has been flagged up for popular discussion since the trials. What it does throw into stark contrast are the writerly assertions with which we started this exploration of island mentality and the hard and sometimes tragic edge of contemporary history.

Established attitude and behaviour patterns are resistant to change, and this is as true on an island as it is on the mainland; and it is probably as true for the monk as it is for the convict, both in their penitentiary cells with plenty of time for self-reflection. One may anticipate the voluntary simplicity of island life, one may anticipate the release from and transcendence beyond an angst-ridden mainland life, and one may hope for the revelation of a self more at ease in the limited demands of the new environment. Despite the dearth of evidence to support optimism, one might be successful, but human nature can be perverse: small island life can just as easily foster negative introspection amongst those so predisposed. The social world of islands is synonymous with being on a boat constantly at sea, governed by storms and tides. But a sailor's well-being is profoundly determined by so many unavoidable interactions with fellow crew members – a reason why crews are so often socially dysfunctional and in a state of near mutiny. Once ashore one seems destined to live with the psychological baggage one has always possessed.

The intimacy that islanders share with their environment can be the envy of visitors. For example, E.J. Banfield, the self-styled "Beachcomber of Dunk Island" off the Queensland coast, surprised visitors with his detailed knowledge of the biological community with which he shared the island – "patting trees as if they were old friends". Here was a man who, according to the London *Times Literary Supplement,* was realising the dream of every boy who has read Ballantyne or Defoe, a dream of being "soaked by the salt sea, drenched by tropical rains, infiltrated by the purifying keenness of tropical light". [24] And these feelings of physical intimacy are echoed by Banfield himself in describing the unique satisfaction inherent in exploring untrodden shores where one might discover how "nature, not under the microscope, behaved". [25]

But whilst Banfield seems at ease with himself in his island environment, the same cannot be said for a multitude of people seeking an island as a means of avoiding or controlling mainland anxieties. Indeed, as Jonah Jones suggests in his sympathetic afterword to Brenda Chamberlain's *Tide-race*, "To seek an island is the wish of those who suffer too deeply from the cut and thrust of mainland life. ... Brenda came to the island [Bardsey] part-wounded in some way, came like a pilgrim searching out healing, hoping almost to master fear. One's first impression of Brenda was of vulnerability. She was small, yet strong of bone with a tall, gothic countenance. But she was susceptible to deep hurt, which she held within." [26]

What Jones could also have added was that there have been many people who have sought out small islands for salvation – whether on tiny Bardsey or in the "paradise islands" of the Pacific – but few seem to have been successful in their quest. And revelation has a tendency to slip from our grasp: even Jean-Jacques Rousseau, who found solace from persecution on the lake island of St Pierre in northern Switzerland, stayed only for two months before throwing himself back into his chaotic mainland life.

The Lighthouse Experience

In seeking the extent to which island life conforms attitudes and behaviour, we can bring our search closer to mainland shores than the far-flung Pacific. Before the automation of the lighthouse system in the British Isles, one way of seeking the apparent benefits of small island life was to become a keeper of the light. Voluntary maroons they were, but they were also paid employees whose work was determined by a strict routine that regulated their duties on the lighthouse and structured their time on and off shore. This schedule could be radically interrupted by sea and weather conditions, but there was always the knowledge that life on the light was shared with a life elsewhere.

It would have been intriguing to investigate whether these men (and the occasional woman) shared an insular mentality that made them particularly suitable for the life, and lighthouse keeping would have been an excellent subject for social psychology research, with its captive population forming an established and reliable study group. However, the Corporation of Trinity House and the Northern Lighthouse Board were notoriously unaccommodating to enquiries that might open up their world to public curiosity; and anyway, the automation of lighthouses means the opportunity has passed.

Research may have thrown up some particular character traits, but it seems that many just followed in their father's footsteps, and were not subject to much in the way of a searching interview. Tony Parker, in his fine book *Lighthouse*, provides us with a series of insightful interview-discussions. As one keeper explains: "How to tell you what it's like? ... Honestly I wouldn't know where to start. It's a different world you see out there, a totally different world; you couldn't imagine it for yourself and I couldn't explain. When you come to think of it, three men living their lives marooned on a rock sticking up out of the sea. Bloody ridiculous isn't it when you come to think about it." [27]

A "tower" is a lighthouse perched on an inaccessible rock that may at times even be semi-submerged. The only surrounding landscape is the sea, and the confines of space would impinge on every second of time. This, and two fellow keepers who were compressed into one's existence, would have determined almost the totality of one's lifeworld. In Parker's accounts of life on a tower, one is characterised by a destructive obsession, and the other reflects on the experience with a sense of repulsion.

In the first interview, "Mary B" describes her husband Simon's first year "wrapped in the lighthouse atmosphere" as being more or less satisfactory for both of them, but she goes on that he then began to change:

" ...it started getting obvious after a while he was obsessed with the place and couldn't wait to go back. I was spending two months thinking about him coming home; then when he did come he spent his entire month ashore thinking about nothing else but going back. I couldn't understand how anyone could be like that, particularly someone who'd been as fond of his home and of me. I never did find out what it was about the place he thought was so marvellous. Time and again I tried to ask him and discuss it, but all he ever said was he felt happy there on the tower in the middle of the sea. It was almost as though he'd got another woman out there; or the lighthouse was another woman he loved more than he did me. ... I can't look on it any other way than it was the life on the tower which changed him.

"Occasionally you see a mention of the place in the

newspapers or hear its name; whenever I do I can't help it, just to hear it makes me shudder. To me it's like some horrible ogre, the place I lost Simon to and my children lost their father. It's almost as if he'd gone there and had an accident and been washed off into the sea or something, and never came back. After he went there he never did come back, I don't know why. I don't like thinking about it."

"Eric G" describes entering "the service" like his father:

"About ninety per cent of people in it are like that; there for want of being able to think of anything better. To be frank you wonder whether there's anything else some of them could do, you can't see them fitting in with any normal job. ...Their system of starting people off...is a bad one, very inefficient. I can't think any modern company would spend time and money on training people without finding out whether they've any aptitude. ... [The keeper] has fifty-six days and nights at a stretch during which he's no contact with the outside world. He doesn't meet anyone apart from the two others he's with. It's as though he's in a monastery or a prison, he's totally shut off from the world. ...I was only staying in for one reason, which was that I wasn't thinking about what I was doing. It's a very insidious process; what happens to you doesn't encourage you to think, and before you know where you are you find you are too old to think any more. ... I began to notice it in time: that when I got ashore I didn't belong in the ordinary everyday world. I'd be talking about things which were real to me, which were what had happened when I was ashore previously. That was eight weeks before, they were gone and forgotten in the minds of everyone else. Or people would be talking to me about something that was going to happen 'next week'; and I'd realise it was meaningless, it had no relevance because 'next week' I was going back on the lighthouse.

"You get very conscious of this. If you went into the pub for a drink you'd deliberately cut yourself off from other people; you'd sit in a corner on your own and wouldn't get into conversation because of letting your ignorance show. People would start to chat with you; they'd mention something that had happened the week before, it might only be something trivial like a film at the local cinema, and you wouldn't even know that it had been showing. It was inevitable they'd ask you if you were a stranger or where you'd been; then when you said you'd been on duty on a lighthouse, they'd start plying you with questions. So no matter which way you turned, you couldn't join in ordinary conversation.

"Another way it used to affect me was in the streets with traffic. It could get frightening because you weren't used to the noise and movement and there were so many people. I was scared crossing the road or going into a large shop. I found it confusing and frightening, and all I wanted to do was get back where I felt safe; get even further away from the world and back to the lighthouse.

"I think that was an appalling state for anyone to be in, let alone a young man as I was then. The existence was totally unreal; it wasn't so much a way of life, it was an escape from life. It's a harmful experience and one people could do well without. I think it very necessary to warn people about it, so they don't drift into it unthinkingly as I did. If you're not prepared to give up your personality to it then I wouldn't recommend it to anybody."

Simon and Eric's experiences working in a tower are generally dissimilar to those described by keepers who worked on larger islands. Indeed, if there is "typicality" in the accounts it is in the matter-of-fact way in which the keepers tell their stories, as if they are more normal than exceptional. What is evident, and hardly surprising, is that condensing one's lifeworld into a tower may squeeze out the possibility of life-enhancing experiences. Otherwise, it seems most likely that lighthouse keepers arrived with a psychological profile that served the needs of the job, and perhaps this was because they tended to come from families where the potential perils of the job were well understood. There is little evidence that the keeper of the light experienced much in the way of transcendental psychological readjustment in the course of his duties.

As a sea kayaker, I've had the good fortune to talk with many keepers working on off-lying islands, including the Farne Islands in Northumberland, Bardsey, Skokholm, the Skerries, South Stack and St Tudwal's in Wales; and the Butt of Lewis and Barra Head in the Outer Hebrides. From these visits I can say nothing more about island mentality other than that the keepers were welcoming and hospitable. I spent two weeks on South Rona in 1964, where the head keeper was clearly entranced by the island, and was delighted to share his island pleasures with young people. His fellow keepers exhibited little curiosity about the island, and seldom left the confines of the light and its buildings.

Islands of the Mind

As it is with lighthouse keepers, so it is with people in general, and our island experience will depend significantly on what we bring to the island. Otherwise small islands are just as likely to pick your foibles down to the bone as they are to wash you in transcendental balm.

Two short stories help to illuminate this last point: W. Somerset Maugham's "German Harry", published in 1924, and D.H. Lawrence's "The Man Who Loved Islands", published in 1926. [28]

In fairness to German Harry, he did not seek an island as an escape from the pain of mainland life. Rather, he was cast away upon an island with a group of fellow seamen, and three years later, when rescue was at hand, he decided to stay – for what turned out to be thirty years, a voluntary maroon on his own. There is a concomitant mystery here, in that during the three years awaiting rescue, eleven of the original sixteen seamen died, and there is the implication that, whatever the circumstances of these deaths, it was enough to persuade German Harry that he had no wish to rejoin a larger human community. This, and the rumour that a collection of magnificent pearls had been stashed away, is the context for our encounter with Harry. Unwelcoming, surly, thankless for his provisions and gifts, and with no apparent interest in the greater world, the narrator is perplexed because "if what they tell us in books were true his long communion with nature and the sea should have taught him many subtle secrets. It hadn't. He was a savage. He was nothing but a narrow, ignorant and cantankerous seafaring man. And then I foresaw the end. One day a pearl fisher would land on the island and German Harry would not be waiting for him, silent and suspicious, at the water's edge. He would go up to the hut and there, lying on the bed, unrecognisable, he would see all that remained of what had once been a man." [29]

This is melancholic enough, but there is an equally grim postscript. In an article in the *Sydney Morning Herald* from November 24, 1951, titled "Grim end to Somerset Maugham's 'German Harry'", John W. Earnshaw claims that "German Harry" is based on a ca. 1922 visit made by Somerset Maugham to Deliverance Island in the Torres Straits, and that the voluntary maroon was in reality the Dane Henry Evolt, who died there in 1928, aged seventy-nine. At the time of Earnshaw's visit to Deliverance Island, about six years after Somerset Maugham, he describes the island as "perhaps half a mile or so in circumference, ringed with a beach of white coral sand, covered with coconut palms dancing in the breeze, and surrounded by a wide fringing reef ... it resembled an island in a boyhood adventure book. ... On the beach we could see no sign of German Harry [Henry Evolt].... So it was with foreboding that we rowed ashore. ... Half under the raised floor, lay what was left of German Harry. Time and hot tropic sun had removed resemblance to a once-sturdy sailor. With outstretched skeletal arms the hermit seemed to clutch at the soil." His right hand and left foot were missing, and the calendar indicated that he had been dead for about two months.

According to Earnshaw, Henry Evolt was born in Denmark in 1849. A seaman, in the 1890s he worked with Louis the Greek on a boat harvesting *bêches de mer* (sea cucumbers) on the coast of Dutch New Guinea, using Deliverance Island as a base. This appears to have been a successful enterprise until the arrival of Joe Austen, or Joseph Augustin de Paoli, who, according to Earnshaw, was a Corsican soldier of fortune who fought in the Crimean and the Franco-Prussian wars. Joining the Communist uprising in Paris in 1871, de Paoli had been arrested and exiled to New Caledonia, but succeeded in escaping from the transport in Melbourne. Then began his wanderings in the Pacific. But his arrival at Deliverance was destined to break up Evolt and the Greek's fifteen-year partnership: when he made off with a boat loaded with turtle shell and other trade goods, and sold the lot on Thursday Island, Louis the Greek left. And Henry Evolt was left with his solitude for the next twenty-eight years. Earnshaw concludes that, "towards his last his greatest fear was that Authority might wrest him from his island refuge and send him to the frightening care of a home for the aged in Brisbane. German Harry was the last of the old New Guinea beachcombers, and his like will not be seen again."

All this seemed credible until I read "French Joe", another of Somerset Maugham's short stories based around Thursday Island in the Torres Straits. Published about two years before Earnshaw's supposed discovery of the demise of Evolt, the story features "Joe", who is described in much the same way as Earnshaw describes Henry Evolt's nemesis – a Corsican soldier of fortune, etc.

Earnshaw claims that Somerset Maugham visited Deliverance Island in about 1922 where he met "German Harry"/Henry Evolt who was the inspiration for his story. "German Harry" was published in 1924. *French Joe* was first published in 1926. And Earnshaw claims to have found Henry Evolt/"German Harry" dead on Deliverance Island in 1928, an event he describes by lifting a character from the story *French Joe*, published two years previously. He makes no mention that he is doing this nor why.

The Island

Certainly the Torres Straits seems to have had more than its share of shipwrecks, and an abundance of lurid tales of the fate of castaways amongst headhunters and cannibals, but "Harry" and "Joe" were far removed from this and were characters derived from Somerset Maugham's experiences there. Indeed, Elizabeth Burchill, in her book *Thursday Island Nurse*, claims that the subject of "French Joe" was in fact a man known as Simon, the oldest man on Thursday Island when she was working there between 1958 and 1960, and it is possible that "Simon" provided the inspiration not only for Somerset Maugham's "French Joe" but for some of "German Harry," too. [30] But this seems unlikely, just as it seems unlikely that Earnshaw was unaware that he was "borrowing" almost line for line a description in one short story to account for what happened in reality to a character in another short story. Earnshaw may be guilty of some sort of mischief, but the convoluted background of fact and fiction may make this difficult to prove.

More certainly – and still in the Torres Straits – a man known locally as Ron Brandt, but whose real name was probably Gosta Brand, lived on his own on Packe Island for nearly twenty years. He was known as the "Swedish Robinson Crusoe", and died in 1981. It is said that locals avoided the island because he had a habit of firing a shotgun at or towards visitors; but in about 1975 Gosta was visited by Torre Zetterlund – a fellow Swede who carried news from Gosta's brother. Zetterlund was captivated to discover everything he had learned to expect of a South Sea island: calm, perfectly blue water barely lapping soft, white sand leading up to sleeping palm trees. Not at all unsociable as the visitor had anticipated, Gosta was clearly affected by the visit and news of his family in a way that led Zetterlund to believe that Brand's life on the island was deeply underscored by loneliness. In departing he wrote, "I had one last look into the cabin of his boat .... There were three guns, two with telescopic sight, a cracked mirror, an old radio, some cans and a pair of old-fashioned spectacles. The sum total of his life, plus loneliness, hardship, and occasional illness. As we left, the outline of where he sat in the boat waving goodbye was getting smaller and smaller. Very soon it would be hard to believe he existed at all." [31]

The narrative of D.H. Lawrence's "The Man Who Loved Islands" is focused on "the man's" sojourns on two islands in the English Channel Isles – Herm and the smaller Jethou – and on one of the much smaller Shiant Islands in the Scottish Hebrides. These islands are in fact un-named in the story, but it is well understood that Lawrence structured his narrative around the author Compton Mackenzie's seemingly obsessive habit of moving from one island to another – from Capri then to Herm, Jethou, the Shiants and Barra. Mackenzie purchased the Shiants in 1925, the main island having been abandoned since 1901 to sheep, a much larger population of rats, and its impressive coastal wildlife. He renovated a cottage and used it occasionally for a day or two to write. Lawrence's ire was directed at the theatrical egotism and self-aggrandisement to which Mackenzie was prone, and when Mackenzie claimed to have fallen in love with the Shiants, it was a good opportunity to prick his obsession with a sharp shot of satire. [32]

As the narrator explains, the man wanted his own island, his very own world. He could not be master of a big island, which was really not much better than a continent. "It has to be really quite small, before it feels like an island; and this story will show how tiny it has to be, before you can presume to fill it with your own personality." Herm is approximately two and a half kilometres long and one kilometre at its widest, and here he introduced his development plan to make it "the Happy Isle ... a minute world of pure perfection, made by man himself", with life anticipated eagerly as "cozy and home-like [where] the little ways and glades were a snow of blackthorn [with] many birds and nests you could peep into, on the island all your own".

But by the third year, "when the wind left off blowing in great gusts and volleys, as at sea, you felt that your island was a universe, infinite and old as the darkness; not an island at all, but an infinite dark world". And this metamorphosis from hope to despair was paralleled by disillusionment in the small social world he hoped to construct as his fellow islanders, who were also employees and servants in his fiefdom, seemed to confound his plans. He commanded deference but inspired neither loyalty nor honesty. Despite considerable expense he was failing to discover happiness and contentment. On the contrary, he was increasingly angst-ridden, and in his frustration he damned the island as "treacherous and cruel. ... In spite of all its fair show of white blossom ... it was your implacable enemy".

Undeterred, and because he was "the man who loved islands", he was not going to be put off by minor setbacks. He moved to a neighbouring small island where he could still be master of his piece of rock in the sea: "like the last point in space ... the island was no longer a world. It was a sort of refuge.". But it was a refuge where even the anticipated pleasure of writing a book on the island flowers became a burden rather than giving him a sense of meaning and purpose. And, with the few employees who had stayed with him from the bigger island understanding "meaning and purpose" even less than he did, the man decided to seek out a more remote, even smaller island.

By the time he arrived on his smallest of islands, like a stowaway in his baggage, he was bringing ashore that "implacable enemy, an infinite dark world" of his own hubris. No island, no place could satisfy his need to make a world that he could fashion exactly to his arrant will. Instead the process of alienation continued remorselessly into paranoia. Now he was even subdued by the idea that rare passing steamers and sailing boats might bring curious people. He learned to hate the human-like sounds emanating from the few island sheep, and to appreciate the absence of any tree that "stood up like people, too asser-tive". And so we leave this man – whose only consolation was knowing that he was alone – "absolutely alone, with the space soaking into him".

The man who loved islands, in his solitude, deserves our compassion. But during his existentially bleak last days on his very smallest island, we witness the lone figure in a fable concerning human arrogance. He believed that on an island he would be able to reconstitute his life away from the distractions of the mainland; that remoteness would not be isolating. The small and the smaller isles were not particularly harsh physical environments, and even whilst railing against the treachery and cruelty of his second island home, he is admitting too "its fair show of white blossom". But his experience is rendered essentially dystopic by trying to impose intractable mainland values, by seeking to possess an island to be his and his only, and where there was no tailing off of his domain into a distance that challenged his authority, where all could be viewed at a glance, then here he could be master of all he saw and touched. Vanity like this can lead to paranoia, and so it did. In the end he was master only of a dead landscape, smothered to death by the neurotic arrogance of his ambition and numbed by the prospect of life itself. [33]

If there is a bitter consolation for "the man" – and perhaps for Compton Mackenzie, too – it may be in the life of D.H. Lawrence after he left England in 1919, an England he found dull and uninspiring, and which he departed in search of a haven that conformed more readily to his own temperament. Germany, Italy, France, North-East Asia, Australia, west coast United States and Mexico all seemed to disappoint him. Like our bleak islander, he lived the life of an exile, exotic but with relationships often more of convenience than challenging the imagination. [34]

Afterword

A Stranger on the Shore

A stranger from the continental interior is visiting the coast for the first time. She enjoyed geography at school, so there is a sense of familiarity as she runs her eye along the shore, noting its beaches, cliffs, capes and bays. But she is disappointed and distracted until, in the distance, fading in and out of view between squalls, she spies an island. She tries to focus her attention because she imagines its outline keeps changing even as it stands firm in a sea of uncertain mood. It seems to draw her in. Its very existence must be precarious, she surmises, and in the fading light she fears it will disappear forever. She decides to find out more.

Our stranger is not alone in having her emotions unsettled by the sight of an island, and there are thousands of them scattered across the oceans, many clustering loosely together as if hoping to avoid the fate of others that are so far-flung they seem to be teetering into infinity. Most of us will never visit them, but those with courage and incentive

have described their encounter with these remote fragments of land most dramatically, their descriptions ranging
from Heaven to Hell and everything in between.

Since the dawn of time, natural philosophers have
searched for evidence to explain their world's place in the
cosmos, just as geographers have sought to transcend a
primitive understanding of the Earth. Empirical evidence
upon which theories are based relied on explorers and
voyagers who, with all manner of motives, saw fit to
make honest reports or to shape accounts for political and
economic advantage. Some sailings had one reason alone to
make these dangerous sorties, and that was to confirm the
belief that there was an earthly paradise and that the "sea to
the west" was the route to the promised land.

But, motive apart, islands confused things by spitting fire
and belching smoke, an acrid stench and horrific sounds
announced their presence in oft-misty weather, and ships
were dragged towards them only to be saved as the island
subsided in a mirage. These scraps of land were discovered
one year and were quite likely to have gone missing the
next. People on the shore might be seen gesticulating
grotesquely, so this was no place to linger. But on other
islands that were about to be "discovered", the people on
the shore could be waving and beckoning invitingly, the sea
lapping the white sand beach, shaded by fruit-laden trees,
with gentle airs transporting the sound of children playing
and the sweet scent of flowers, and all bathed in the pure
light of the sun.

Whether the crew discovered paradise or purgatory, or
whether their ships were wrecked and they were cast away,
stranded and marooned, those fortunate enough to make

it back home safely would harbour strange, exotic tales
upon which storytellers could embellish. Out of this came
Robinson Crusoe and a host of would-be voluntary maroons
wishing to emulate him.

Our imagination is stoked by islands that, in birth as
in death, seem to defy conventions of time. If "time waits
for no man" and "~~no~~ every man is an island" then islands
may perhaps be excused for being immature and flighty
– thrown up overnight in the "ring of fire" and destroyed
next day by tsunami, ocean and estuarine currents, swollen
rivers and king tides.

A fragile geography, then, and one where indigenes
survive only by dint of remoteness, and where "discovery"
has fatal consequences when "primitive" is just a stain
to be rubbed out. With their unique adaptations, biotic
communities seem to get along for thousands of years too
until their "closed systems" are penetrated, whereupon their
innocence is exposed.

Modest by nature, islanders seek a modicum of stability
almost always against the background of limited resources.
They are good at making the most of small-scale traditional
activities. But they have learned to exploit the sometimes
exotic beauty and rather particular history of their home,
so becoming co-conspirators in presenting their reality as
paradise. And they have discovered the double-edged sword
of tourism, which has reminded them of their fragility and
vulnerability.

From time immemorial, the sea and its sky above have
provided throughways for nations to pursue their ambitions. Small, remote islands find themselves involved in this
enterprise more often than they might like, and so run the

risk of becoming pawns in the power games of empire, annexed and traded like disposable and exchangeable units. Whether it is for something as seemingly innocent as guano, or for something as evil as the testing of atomic bombs, at worst islands can be exploited virtually out of existence. People living through this are certainly on the other side of paradise.

★

Our stranger is of a quizzical disposition, determined and with a wayward curiosity. She returns to the same coast, now feeling like something of an expert on the subject of small islands. But she has never set foot on one, and this is something she wants rectified. She seeks a ferryman, but none seem willing.

Her ambitions frustrated by what seems like a mere twist of fate, she sits overlooking *her* island until she gradually feels more at ease. Perhaps it is all for the best that her dream is unattainable. She can now envision what goes on there, and needs no "reality" to interfere with such rich imaginings.

Island Tales

In February 2010, on Robinson Crusoe Island near South America, a twelve-year-old girl felt the ground tremble and noticed the lobster boats in the harbour were becoming increasingly agitated and the sea more and more rough. She raised the alarm, using the town hall bell, rousing most of the island families who were at siesta. The islanders followed evacuation signs set up for emergencies and fled to higher ground.

It was reported that just three minutes later huge waves crashed on to the land, sweeping 300 metres into the village of 650 souls as the sea rose twenty metres. The school, the pre-school, the municipal offices, the church, the cultural centre, most of the shops and a number of homes were seriously damaged. But because of the action of one young girl nobody lost their lives. [1]

Kanton Island, part of the island state of Kiribati, is a fifteen-kilometre-long coral strip enclosing a lagoon. Before World

War II it was a stopping-off point for the Pacific Clipper flying boat, and as such Britain and the United States vied for control of the island. During the war, it was occupied by 1,200 troops as a base from which to attack the Japanese. The airlines pulled out in the 1960s, and it was briefly a US missile tracking station before the British and United States abandoned the atoll. Even the post office closed in 1976.

In May 2010, a British sailor delivering a sail boat from Hawaii to Brisbane found the population of fourteen adults and ten children – the guardians of this World Heritage Site – in desperate need of food. Their supply vessel had been delayed three months, with little chance of it arriving for another month. The sailor helped the islanders with some provisions and used his satellite telephone to contact the British coastguard in Falmouth, 15,000 kilometres away, who helped coordinate a food drop. [2]

According to a BBC news report, in 1993 the Royal Air Force strayed from its firing range and bombed the wrong island off the north coast of Scotland.

With its 1,600 kilometres of coastline, the longest in Europe, Greece has nearly 10,000 islands and islets. For the many inhabited islands, ferry services are vital, and popular routes can be lucrative for ferry operators. It seems that in the late 1990s furious fights over operating licences led to ships being involved in near-misses, with captains trying to subvert timetables by squeezing into small island ports. Even family rivalries reached fever pitch with the deliberate ramming of ferries, and one brother allegedly tried sinking his sibling's ship by hiring divers to rig its hull with explosives. [3]

It is rumoured that a firm of London estate agents once purchased an estate on the Isle of Lewis and – in keeping with the history of landlordism in much of the Scottish Highlands and Islands – they were intent on limiting expenditure and turning a profit. They reportedly sold tiny sea rocks and islets to buyers in the south of England, and in so doing they made tens of thousands of pounds, more than they had paid for the whole estate.

In 1856, the population of the island of Pitcairn had risen rapidly to 187, precipitating fears that the land and the diminishing fishing would no longer be able to sustain them. An evacuation was organised to Norfolk Island, a former penal colony. But many of those who left soon became dissatisfied, and two years later they sailed back to Pitcairn. As they approached the island "capital", Adamstown, they noticed that their homes were not derelict and overgrown. Some of the buildings had evidently been lived in, whilst others had been dismantled piece by piece. A message was scratched on a slate signed by Captain Josiah Nickerson Knowles.

In February 1858, his vessel, *The Wildwave*, with a crew of thirty and ten passengers, was cast ashore on the reef of nearby Oeno Island. They were fortunate to find a source of brackish water, which they shared with a multitude of land crabs and rats. Salvaging as much as possible from the wreck, they set up camp.

With six men, Knowles sailed for Pitcairn in a small boat, and after two days they were able to make a difficult landing, but their boat was eventually crushed by waves. Discovering Adamstown abandoned, Captain Knowles was disappointed to discover that there were no whaling vessels in the area.

They reluctantly set about disassembling some of the homes to help in the construction of a boat, and after four months they set sail. A twelve-day voyage brought them to Nuku Hiva Island in the Marquesas, where it was their good fortune to discover an American man-of-war. From there, Knowles was able to effect the rescue of crew members left behind on Pitcairn and Oeno.

Fourteen years after his ordeal, the captain returned to Pitcairn, where he was feted by the inhabitants. [4]

A British couple, Taffy and Bonnie Bufton, moved to the two-square-kilometre Guiana Island, just 100 metres off the north-east coast of Antigua, in 1965. They farmed the land for the owner, a London lawyer, and when he died in 1972, the couple claimed that they had been promised a ninety-nine-year lease. The agreement was never formalised, but they remained as sole inhabitants of the island, taking pride in caring for the wildlife, which included the endangered West Indian Whistling Duck amongst other rare fauna and flora.

That was until 1997 when the Antiguan government and a billionaire Malaysian businessman announced a $600 million Asian Village development for the tiny island and the adjacent coastal strip, including a 100-room hotel, a golf course, casino, shopping village, conference centre, theatre and water theme park. If the couple continued to resist eviction, then the government had passed the Taffy and Bonnie Act specifically for the eventuality, which allowed for a £10,000 fine or a nine-month prison term to be imposed on the Buftons and those campaigning with them. It also provided them with a seafront home on Antigua and a $567 a month stipend. Passions ran so high that the police claimed that Taffy shot and wounded the Prime Minister's brother and himself in a fracas. [5]

In 2009, the television psychic Uri Geller purchased Lamb Island, in Scotland's Firth of Forth. It is a jagged volcanic outcrop measuring about 100 metres by fifty metres, for which he is reputed to have paid £30,000. This is probably cheap for an island whereon he has identified links as diverse as the pyramids, King Arthur, King Robert the Bruce and the ancient kings of Ireland.

The Island

In 2005, *The Guardian* newspaper reported that a group of "adventurers" had been active on Robinson Crusoe Island. With the aid of a mini robot able to scan fifty metres deep into the earth, the article stated, they unearthed an estimated 600 barrels of gold coins and Inca jewels, valued at $10 billion. Treasure was reputedly buried there in 1715 by the Spanish sailor Juan Esteban Ubilla y Echeverría, and the rumour has long occupied the imagination of treasure hunters. The finders claimed half the treasure was theirs, and said they would donate it to non-profit-making organisations. The Chilean government said that the finders were not entitled to a share of the treasure.

It transpired that there was no entitlement to treasure because the treasure hunters, like so many fortune seekers throughout history, appear to have found nothing. But they were not giving up, and six years later *The Santiago Times* reported that Bernard Keiser – an American millionaire who had made his money supplying material for space suits to NASA, and who had already spent $2 million on the project – planned to search the island for a fifth time. [6]

In September 1988, *The Observer* newspaper published an article titled, "Former Surrey Sewage Clerk leads Christmas

Island in Rebellion". According to this article, "a remote island in the Indian Ocean, with an Asian community led by an Englishman with the unlikely name of Gordon Bennett, is locked in battle with the Australian Government over its survival. The 900 Chinese and Malays, led by the forty-four-year-old ex-sewage clerk, called 'Lord Jim' by the Australian administrators, are this month taking the Australian Government to court over their rights on Christmas Island, more than a thousand miles from the Australian mainland. A twelve-year feud with the administration has led to Canberra closing down the island's mine and shipping all the workers out, amidst accusations of racial discrimination and political trickery." [7]

In September 1967, Major Roy Bates occupied a derelict steel and concrete World War II defence platform ten kilometres off the Essex coast in England. He declared it the Principality of Sealand. Bates gave his artificial island state its own flag, constitution, passport, postage stamp, currency and national anthem.

In 2012, with the island state itinerant population of four supplemented by some skilled professional players who saw this as an opportunity to become "internationals", Sealand's side played its first football match. This took place on the mainland to save too many lost balls "overboard". They lost 3–1 to the Chagos Islands, a team living in exile since the islands were evacuated by the British government forty years previously to make way for a US military base (see Chapter Six). [8]

Henderson Island lies some 200 kilometres from Pitcairn Island, 6,500 kilometres from Panama and 2,150 kilometres from Tahiti. When Pitcairners first visited it in 1851, looking for improved fishing grounds and supplies of timber, they found eight human skeletons lying together.

With its covering of scrubby bushes and with only a meagre source of brackish water, it was decided to leave the island to the skeletons. [9]

In 1777, a cask came ashore on the Pembrokeshire coast of Wales. On the outside were painted the words "Open This And You Will Find A Letter", and inside the note read: "Fear we shall all perish. Our water nearly all gone, our fire quite gone out, and our house in a quite melancholy manner ... we could not have kept the light above 16 nights longer for want of oil and water." It had drifted in two days from the Smalls Islands lighthouse, which at the time was occupied for repair by a blacksmith and the light's designer whose engineering expertise was hitherto as a distinguished maker of musical instruments. [10]

A story circulates on Barra, in the Outer Hebrides, concerning Mingulay twelve miles to the south. Having received no news from the island for several months, a sailor landed there and found everybody dead, apparently from the plague. Fearing contamination, his fellow crew members made rapid departure, leaving him marooned on the island. He was not relieved of his solitary condition for a year and a day. [11]

In 1835, the Mexican government hoped to extract high profits from otter hunting on San Nicolás in the Californian Channel Islands, and so the indigenous population was evacuated. In their haste, they left behind a woman who had walked inland to seek her child. Subsequently, she was occasionally sighted by sailors and hunters, but she always fled when people approached. In 1853, she was taken to the mission in Santa Barbara, where she died within a few days. [12]

In February 1979, five men left Maui in Hawaii for a fishing trip in their five-metre skiff, named *Sarah Joe*. Surprised by a storm, the crew was reported missing, and the coastguard and a group of local people – including marine biologist John Naughton – searched long and hard, but no trace of the *Sarah Joe* or the five men was found.

Nine years later, Naughton was on a wildlife expedition to a deserted atoll called Taongi, part of the Marshall Islands and approximately 3,750 kilometres from Maui. There on a beach he discovered a small boat with a Hawaii registration mark. Several feet away there was a pile of rocks – apparently a shallow grave with a human jawbone protruding. In the grave was an unbound stack of papers measuring about fifteen centimetres square, each page separated from the next by a slip of tin foil.

The coastguard linked the boat to the *Sarah Joe* and dental records proved the jawbone belonged to Scott Moorman – one of the missing *Sarah Joe* fishermen. Later, a few more of Scott Moorman's bones were found in the same area, and divers found the boat's engine wedged underwater in nearby coral. [13]

According to a popular seventeenth-century chronicle, French pirates captured a passenger ship en route from England to Ireland in 1615. The ship was taken in tow with three passengers left on board. A violent storm blew up, forcing the pirates to cast off their prize, which was swept far out into open sea. The unfortunate "crew" of the ship had no water, and their only food was a bag of sugar, and they suffered great hardship until they were flung upon a small island as their ship broke up around them. One person was drowned.

Cast upon an isle not much more than a large rock devoid of all vegetation, there was no water save that which collected in rocky pockets as salty rainwater. If they had any luck at all, it was in three slabs of rock forming a kind of bench, which had perhaps been used by fishermen to dry fish. There was also a board salvaged from the ship, and with this and the slabs they made a shelter. They also had a knife, and were able to catch seals and gulls, and there were eggs, too.

After six weeks, the Poor Englishman – for that was his name – woke to find his fellow castaway had vanished. Now he had no friend with whom to divide his terrible ordeal. And soon after, he lost the knife, which he replaced by sharpening a nail. With winter fast approaching, and wearing nothing but rags, his options seemed to be either to die of cold or to starve. Needing to spend more time in his shelter, he was resourceful as ever: he baited a sliver of wood with a smear of seal fat, stuck it through a crack in the wall of his shelter, and when a bird landed on it he somehow grabbed it.

Nearly a year after he had become the sole occupant of a rock in the ocean, a cargo ship from Norway was becalmed and swept by tide and currents towards his island. The crew used a skiff to tow the ship out of danger, and took the opportunity to come ashore in search of birds' eggs. Seeing a ghostly figure in the distance, and fearing the worst, they fled back to their skiff and rowed energetically back to their ship.

But – at last – the Poor Englishman was rewarded for his endurance as the ship once again drifted dangerously toward the shore. This time the crew got a better look at him, still more a spectre than a human being, with its near-naked, black, hairy body and emaciated face hollowed to distorted eyes. But he was rescued, and later was able to tell his tale, over and over, when he made it back to his home town in Ireland. [14]

The military junta of Augusto Pinochet established an infamous concentration camp on Dawson Island off Punta Arenas in the Straits of Magellan. It was built on the ruins of the last Silesian mission to Tierra del Fuego, where well-meaning priests presided over the extinction of the island's Selknam people.

It was reported in 1988 that, feeling increasingly beleaguered, Pinochet was preparing a bolthole for himself on Robinson Crusoe Island, the home of the maroon Alexander Selkirk in the early eighteenth century. [15]

Author's Note

Throughout the 1960s, I spent my summers on small islands, on what were rightly called "expeditions" run by the excellent Schools Hebridean Society. With people qualified in every way that matters – with intelligence, curiosity, imagination and enthusiasm – I explored Gometra, South Rona, Harris, South Uist and several more islands along the way. As I recall, our time was spent finding out about "our" island, simply getting to know what was where and why. These forays stimulated my curiosity for further study so that for my undergraduate degree I did fieldwork on Scarp and Scalpay off Harris, an enquiry into depopulation from 1745. I still have a washed-out black and white photograph of a "field" of oats so small that I used my shoe to indicate scale. By the end of 1971, Scarp joined the long list of depopulated islands in the Hebrides.

In between exploring Hebridean islands, I journeyed to Corsica, Sardinia, Lipari, Stromboli, Vulcano and Sicily; and the 1960s closed with a student expedition to Carriacou in the Grenadines, where we undertook land-use mapping for six weeks. Those were the days when a pair of boots, a rucksack and a map – and the curiosity to use them – were *de rigeur* for geographers to pursue their trade.

In the early 1970s, working in North Wales, I kayaked about Anglesey, working the tides in the Menai Straits and around South Stack, the Skerries and Bardsey; then further afield to Skomer and Skokholme, and to the Farne Islands. By the middle of the decade, I was back in the Hebrides, culminating, in the wonderful summer of '76, with a circumnavigation of the Outer Hebrides by kayak.

This was a springboard to visit more distant shores – the islands south of the Beagle Channel down to Cape Horn in 1977–78 and the Alaska Panhandle from Prince Rupert to Sitka a year later. Passing through British Columbia on the way to embarking upon this last trip, I laid plans that would enable me to spend the first half of the 1980s at the University of Victoria on Vancouver Island, and so to explore by kayak and sailboat the myriad of islands that adorn this Pacific coast. In a parallel life, I completed a doctoral dissertation – about islands.

Back in Europe, I explored lesser-known islands in the Mediterranean under sail and then crossed the Atlantic, mooring in the Canaries, Cape Verde and Fernando de Noronha off north-east Brazil.

A *visitor* may claim to have been to an island by stepping ashore, even for a moment. Time is often in short supply and their *trip* is brief by necessity. For me, claiming to be a *traveller* means that things are a bit more complicated, and "being there" means staying overnight on, moored to, or at the very least closely anchored off the piece of land. At last count, I have claimed 153 islands.

Author's Note

Looking back, I realise this was done with fairly modest ability, a degree of somewhat selfish perseverance, and with supportive companions willing to push and shove where necessary in order to share the joys of overcoming strange challenges. And the islomania? Well, in truth, perhaps it has just crept up on me to become the kind of affliction that demands the writing of a book!

Acknowledgements

Very long ago, on the shores of Vancouver Island, I was fortunate to have at my disposal the considerable resources of the University of Victoria. This enabled me to undertake research that forms a key foundation stone for this book, and I owe students, staff and faculty a debt of gratitude for their support and encouragement.

Also in Canada, Dave Robinson, Lorne MacIntosh, Marilyn Bowering and Michael Elcock, and the Institute of Ocean Sciences in Sidney (British Columbia) have helped me in numerous and various ways.

In the United States, Ray Bradley has kept me right on global warming but, as with all else in the text, I alone can be held responsible.

For guidance, inspiration, and for encouragement after reading bits and pieces of early drafts I wish to thank Margaret Kellas and Gunars Libeks, Javier Alli Vienge, the late Frank Goodman, Richard Light, Aly Kellas, the late G.W.S. (Billy) Robinson, Steve Royle, Andrew McNeillie, and Robbie Nicol and a host of student friends. I am too easily diverted from the numerous challenges of writing, and too prone to delay until I am in "the right place". In helping to overcome these self-imposed handicaps, I am indebted to Raphael Rassié (for Arrous), and to Mary Macleod (for Crovie).

Uta Rach and Koit Teng ensured there was a loving home on Vancouver Island from which to sail forth on numerous island adventures. I am particularly indebted to Sarah Manning and to Jim Perrin for advice and encouragement and for making the link to my publisher, Saraband. Without the gentle wisdom of Doug Porteous the project would have foundered upon an island's rocky shore, far away and long ago. I was fortunate indeed to have Sara Hunt and Craig Hillsley of Saraband as expert guides for the important last leg of the voyage. Pete and Marianne Smith of Picturemaps worked tirelessly and, despite my uncertain demands, provided excellent illustrations for the text and cover.

I owe so much to my parents, who encouraged my adventurous island instincts. My only regret is that Christine Smith is not here to share this book with me. She never lost faith even when I faltered. And lastly, I must pay tribute to Caroline with whom I have shared so many adventures, and whose love ensured the boat docked safely on that far island shore.

References

Chapter One: Lost in Space, Lost in Time

I am particularly indebted to Geoffrey Ashe's *Land to the West* (1962) for revealing the significance of Saint Brendan's *Navigatio* and for his interpretation of the early Irish explorations into the western Atlantic. For island themes in imaginative literature, and for the island stories in *Gulliver's Travels*, I have followed the lead of Manguel and Guadalupi in *The Dictionary of Imaginary Places* (1980) and J.S. Bowman in *A Book of Islands* (1971). Henry Stommel's *Lost Islands: The Story of Islands that have Vanished from Nautical Charts* (1984) is a fascinating account of imaginary islands discovered, undiscovered, rediscovered and eliminated. D.S. Johnson's *Phantom Islands of the Atlantic* (1994) introduced me to seven charted islands that never were.

1. Betts, R. (1986). *The Allure of the Island: The Western Imagination and the Search for New Space*. Paper presented at the Islands '86 Conference of the Islands of the World. University of Victoria, BC, Canada.

    Brookfield, H.C. (1990). *An Approach to Islands*. In W. Beller, P. D'Ayala & P. Hein (Eds.), *Sustainable Development and Environmental Development of Small Islands*. Man and the Biosphere Series, Vol. 5. Paris and Carnforth, U.K.: UNESCO/Parthenon.

    Cook, S. (1998, October 24). Drops in the ocean. *The Guardian*, pp. 10–11.

    Small Island Developing States. (2013, August). United Nations – OHRLLS. Retrieved from http://unohrlls.org/

custom-content/uploads/2013/08/SIDS-Small-Islands-Bigger-
Stakes.pdf

2. Meinig, D.W. (1979). *The Beholding Eye: Ten Versions of the Same
   Scene*. In D. Meinig (Ed.), *The Interpretation of Ordinary Landscapes*
   (pp.33–48). New York: Oxford University Press.

3. Sullivan, R., Mason, R. & Payne, C. (2014). *North Brotherton Island:
   The Last Unknown Place in New York City*. New York: Fordham
   University Press.

4. **Rockall**
   Fisher, J. (1956, p.12). *Rockall*. London: Geoffrey Bles.
   Macintosh, J.A. (1946, subtitle). *Rockall*. Oban, U.K.: Hugh MacDonald.
   St. Kilda
   Thomas, L. (1983, p.131). *A World of Islands*. London: Michael
   Joseph/Rainbird.
   North Rona
   Fraser Darling, F. (1952, p.144). *Island Years*. London: G. Bell
   (Readers Union Edition).
   The Shiant Islands
   Nicolson, A. (2002, p.141). *Sea Room*. London: Harper Collins.

5. **Tristan da Cunha**
   Munch, P. (1971, p.2, 6). *Crisis in Utopia; The Ordeal of Tristan da
   Cunha*. New York: Thomas Y. Cromwell
   Nam
   Ellis, W.S. & Blair, J.P. (1986, p.825). Bikini – A Way of Life Lost.
   National Geographic, 169, pp.810–834.
   Easter Island
   Doug Porteous, in discussion.
   Bouvet Island
   Ramsay, R.H. (1972, p.256). *No Longer on the Map – Discovering
   Places that Never Were*. New York: Viking.

6. Lauzon, M., Lauzon, L., Virgo, S. & Adalian, Y. (1974, n.p.). *Verse,
   Wind Song and Equinox*. Port Clements, BC: Catspaw.

7. Durrell, L. (1960, p.15). *Reflections on a Marine Venus*. New York:
   Dutton.

8. Conover, D. (1980, pp.110–111). *One Man's Island*. Markham, ON:
   Paperjacks.

9. Engel, M. & Kraulis, J.A. (1981, p.56). *The Islands of Canada*.
   Edmonton, AB: Hurtig.
   Michener, J.A. (1967). *Tales of the South Pacific*. New York: Bantam.
   Thomas, L. (1983). *A World of Islands*. London: Michael Joseph/
   Rainbird.
   Royle, S. (2014). *Islands: Nature and Culture*. London: Reaktion.

10. Ashe, G. (1962). *Land to the West*. London: Collins.
    Babcock, W.H. (1922). *Legendary Islands of the Atlantic: A Study in
    Medieval Geography*. New York: American Geographical Society.

References

Bowman, J.S. (1971). *A Book of Islands*. New York: Doubleday.

De Camp, L.S. (1970). *Lost Continents: The Atlantis Theme in History, Science and Literature*. New York: Dover.

Manguel, A. & Guadalupi, G. (1980). *The Dictionary of Imaginary Places*. New York: Macmillan.

Palmer R. & Cuffari, R. (1975). *A Dictionary of Imaginary Places*. New York: Henry Z. Walck.

11. Ashe (1962).

12. Cited in Firestone, C.B. (1924, p.270). *The Coasts of Illusion: A Study of Travel Tales*. New York: Harper.

13. Johnson, D.S. (1994). *Phantom Islands of the Atlantic*. Fredericton, NB: Goose Lane.

14. Mandeville, Sir J. (1983 ed.). *The Travels of Sir John Mandeville* (C.W.R.D. Moseley, Trans.). Harmondsworth, U.K.: Penguin.

    Manguel & Guadalupi (1980).

15. Rabelais, F. (1944 ed.). *The Complete Works of Rabelais*. (J. Le Clercq, Trans.). New York: Random House.

16. Manguel & Guadalupi (1980).

17. Bowman (1971).

18. Daws, G. (1968). *Shoal of Time: A History of the Hawaiian Islands*. New York: Macmillan.

    Higginson, T.W. (1898). *Tales of the Enchanted Isles of the Atlantic*. Great Neck, NY: Core Collection.

    Wellard, J. (1975). *The Search for Lost Worlds*. London: Pan.

19. Ashe (1962).

    Babcock (1992).

    Higginson (1898).

    Johnson (1994).

    Morison, S.E. (1971). *The European Discovery of America; The Northern Voyages AD 500–1600*. New York: Oxford University Press.

    Sauer, C.O. (1968). *Northern Mists*. Berkeley and Los Angeles, CA: University of California Press.

    Wilson, B. (1998). *Dances with Waves*. Dublin: The O'Brien Press.

20. O'Siochain, P.A. (1962, p.55). *Aran – Islands of Legend*. Dublin: Foilsiuchain Eireann.

21. Morison (1971).

    Stommel, H. (1984). *Lost Islands. The Story of Islands That Have Vanished from Nautical Charts*. Vancouver, BC: University of British Columbia Press.

    Westropp, T.J. (1912). Brazil and Legendary Islands of the North Atlantic. *Proceedings, Royal Irish Academy*, Third Series, 30, (Section C), pp.223–260.

22. Sayers, P. (1973, p.178). *Peig: The Autobiography of Peig Sayers of The Great Blasket Island*. (B. MacMahon, Trans.). Dublin: Talbot.

23. Eliade, M. (1961, pp.61, 12). *Images and Symbols*. London: Harvill.
24. Ashe (1962, p.49).
25. Johnson (1994).
    Kemp, P. (Ed.). (1976). *The Oxford Companion to Ships and the Sea*. London: Oxford University Press.
    Stommel (1984).
26. Fleming, F. (1999, p.317). *Barrow's Boys*. London: Granta.
27. Boorstin, D. (1983). *The Discoverers*. New York: Random House.
    Daws (1968).
    Sauer (1968).
    Stommel (1984).
    Wagner, H.R. (1933). *Apocryphal Voyages to the North-West Coast of America*. Worcester, MA: Antiquarian Society.
28. Corliss, W.R. (1983). *Handbook of Unusual Natural Phenomena*. Garden City, NY: Anchor Press/Doubleday.
29. Stommel (1984, p.34).
30. Birkett, D. (1998). *Serpent in Paradise*. London: Picador.
31. Stommel (1984).
32. Marks, K. (2012, November 23). Undiscovered country: maps delete island that never was. *The Independent*, p.33.
33. Stommel (1984, p.104).
34. Wiggins, M. (1989, p.63). *John Dollar*. Harmondsworth, U.K.: Penguin.

Chapter Two: Castaways, Maroons & Beachcombers

Daniel Defoe, Michel Tournier and Tom Neale are the inspirations for this chapter. I have drawn on James Simmons (1980, 1998) for his accounts of the life of Tom Neale on Suvarov. Walter de la Mare's *Desert Islands and Robinson Crusoe* (1930) has been a constant companion.

1. Lichfield, J. (2007, February 5). Slavery and its legacy. *The Independent*, pp.24–25.
2. Leslie, E.E. (1988). *Desperate Journeys, Abandoned Souls*. Boston, MA: Houghton Mifflin.
3. Moore, J.R. (1958, p.268). *Daniel Defoe – Citizen of the Modern World*. Chicago: University of Chicago Press.
   De la Mare, W. (1930). *Desert Islands and Robinson Crusoe*. London: Faber & Faber.
   Gove, P.B. (1961). *The Imaginary Voyage in Prose Fiction*. London: Holland.
   Royle, S. (2014). *Islands: Nature and Culture*. London: Reaktion.
   Swados, H. (1961). *Afterword*. In D. Defoe, *Robinson Crusoe*. New York: New American Library (Signet Classic Edition).

References

4. Defoe, D. (1972 ed.). *Robinson Crusoe*. London: Oxford University Press.

5. Forster, E.M. (1962). *Aspects of the Novel*. Harmondsworth, U.K.: Penguin.

6. Gove (1961).

7. De la Mare (1930, p.19).

8. Bowman, J.F. (1853, pp.4–5). *Island Home; or The Young Castaways*. London: Nelson.

9. Tournier, M. (1969). *Friday, or, The Other Island*. (Norman Denny, Trans.). Garden City, NY: Doubleday.

10. Cited in Simmons, J.C. (1998, p.16). *Castaway in Paradise*. Lanham, MD: Sheridan House.
    Steele, R. (1713, December 3). Alexander Selkirk. *The Englishman*, 16.

11. Oliver, D.L. (1962, p.108). *The Pacific Islands*. Cambridge, MA: Harvard University Press.

12. Maude, H. (1968). *Of Islands and Men*. Melbourne: Oxford University Press.
    Michener, J.A. & Grove Day, A. (1957). *Rascals in Paradise*. New York: Random House.
    Osbourne, L. (1921). "A son of empire". In *Wild Justice; Stories of the South Seas*. New York: D. Appleton.

13. Noonan, M. (1983). *A Different Drummer. The Story of E.J. Banfield, The Beachcomber of Dunk Island*. St Lucia, Australia: University of Queensland Press.

14. Sturges, F. (2012, January 26, p.43). Still lapping up the delights of a sure-fire success. *The Independent*, p.43.

15. Kingsland, G. (1984, p.52). *The Islander: The Man Who Wanted to Be Robinson Crusoe*. London: New English Library.

16. Neale, T. (1966). *An Island to Oneself*. London: Collins.

17. De Ridder, J. (1984, October). Remembering Tom Neale – The Hermit of Suvarov. *World Cruising*, p.147.
    Frisbie, R.D. (1944). *The Island of Desire*. Garden City, NY: Doubleday, Doran.
    Rockefeller Jr, J.S. (1957). *Man on His Island*. New York: W.W. Norton.
    Simmons, J.C. (1980, November). The Hermit of Suwarrow. *Oceans*, pp.3–6.
    Simmons, J.C. (1998). *Castaway in Paradise*. Lanham, MD: Sheridan House.

18. Cited in Simmons (1980, p.3).

19. Simmons (1980).

20. Stommel, H. (1984, xv). *Lost Islands. The Story of Islands That Have Vanished from Nautical Charts*. Vancouver, BC: University of British Columbia Press.

21. Simmons (1980).
22. Simmons (1998, p.229).
23. De Ridder (1984).
    Thompson, B. (1976a, January). Suwarrow Atoll. *Pacific Skipper*, pp.31–35.
    Thompson, B. (1976b, May). Sail to Suwarrow: In search of a hermit. *Better Boating*, pp.24–28.
24. An Island to Oneself (website). Retrieved from http://www.riverbendnelligen.com/books.html
25. Neale (1966, x).
    Rockefeller Jr. (1957, p.214).

Chapter Three: A Fragile Geography

For Mont Saint Michel and "L'Île Mysterieuse" I have followed John Lichfield's thought-provoking accounts in *The Independent* newspaper. For an appreciation of the related problems of fragile environment and economic development I am indebted to Stephen Bass and Barry Dalal-Clayton for *Small Island States and Sustainable Development* (1995). MacArthur and Wilson's classic *The Theory of Island Biogeography* (1967) remains a valuable aid to understanding the vulnerability of small, remote island ecosystems. For the interlinked problems of ecology, society and economy in the Galapagos, I am particularly indebted to Godfrey Merlen, and to the work of Lu, Valdiva and Wolford (2013). Charles MacLean's *Island on the Edge of the World* (1972) is a sympathetic and well-considered account of St Kilda as a microcosm of social, economic and cultural processes.

1. Somerset Maugham, W. (1954). *The World Over, The Collected Stories, Volume 2*. London: Heineman (Reprint Society Edition).
2. Simkin, T., Siebert, L., McClelland, L., Bridge, D., Newhall, C., & Latter, J. H. (1981). *Volcanoes of the World*. Stroudsburg, PA: Smithsonian Institution/Hutchinson Ross.
3. Lichfield, J. (2010, August 14). The island that came out of nowhere. *The Independent*, pp.30–31.
4. Lichfield, J. (2008, May 31). Battle for Mont St Michel erupts. *The Independent*, p.31.
   Lichfield, J. (2013a, July 26). Abbey happy after being cut off for the first time in 134 years. *The Independent*, p.30.
   Ferrer, S. (2006, July). 150 million euro 6-year desilting scheme. *The Connexion*, p.5.
5. Ramsay, R.H. (1972). *No Longer on the Map – Discovering Places that Never Were*. New York: Viking.

References

6. Bass, S. & Dalal-Clayton, B. (1995). *Small Island States and Sustainable Development: Strategic Issues and Experience*. London: Environmental Planning Group, International Institute of Environment and Development, Environmental Planning Issues No. 8.

7. Drummond, T. (1997, September 1). Under the volcano. *Time*, pp.24–25.

   Hillmore, P. (1989, September 24). A Caribbean Eden picks up the pieces. *The Observer*, p.11.

   White, M. (1997, August 25). Pleas for Montserrat get Short shrift. *The Guardian*, p.3.

8. Brown, D. & Mahmud, A. (1991, May 6). Now Bangladesh needs a wave of aid. *The Guardian*, p.7.

   Buncombe, A. (2008a, February 7). World's largest river island washing away under flood. *The Independent*, p.37.

9. Karmalkar, A.V. & Bradley, R.S. (2017). Consequences of global warming of 1.5 degree Celsius and 2 degree Celsius for regional temperature and precipitation changes in the contiguous United States. *PLOS ONE*, 2(1).

10. Bass *et al.* (1995).

    Caramel, L. (2014, July 1). Besieged by the rising tides of climate change. *The Guardian*. Retrieved from https://www.theguardian.com/environment/2014/jul/01/kiribati-climate-change-fiji-vanua-levu

11. Buncombe, A. (2008b, July 15). Trouble in paradise. *The Independent*, pp.24–25.

    Byrnes, S. (2012, February 12). Are we really going to abandon the P.M.'s new best friend? *The Independent on Sunday*, pp.10–11.

    Lean, G. (1990, December 16). The disappearing islands. *The Observer*, p.37.

12. Marks, K. (2007a, July 16). A country fights against the tide. *The Independent*, pp.1–3.

13. Carlquist, S. (1965). *Island Life: A Natural History of the Islands of the World*. Garden City, N.Y.: The Natural History Press.

    Conklin, M. (1985). *Islands in Time*. Camden, ME: Down East.

    MacArthur, R.H. & Wilson, E.O. (1967). *The Theory of Island Biogeography*. Princeton, NJ: Princeton University Press.

    Power, D.M. (Ed. 1980). *The California Islands*. Santa Barbara, CA: Museum of Natural History.

14. Royle, S. (2014). *Islands: Nature and Culture*. London: Reaktion.

15. Labrecque, M., Ouimet, J. & Clua, E. *Big Migrations Expedition Clipperton 2016*. Retrieved from www.diveclipperton.n2pix.com/pdf and http://www.diveclipperton.n2pix.com/pdf/clipperton-expedition-2016-report-explorer-club.pdf

16. Labrecque *et al.* (2016, p.25).

17. Bryan, E.H. (1963). *Discussion*. In F.R. Fosberg (Ed.), *Man's Place in the Island Ecosystem*. Hawaii: Bishop Museum Press.

Golson, J. (1972). *The Pacific Islands and Their Prehistoric Inhabitants*. In R.G. Ward (Ed.), *Man in the Pacific Islands* (pp. 5–33). Oxford: Clarendon Press.

MacLean, C. (1972). *Island on the Edge of the World; Utopian St Kilda and its Passing.* London: Tom Stacey.

Stoddart, D.R. (1987). *On Geography and its History*. Oxford: Blackwells.

18. Cousteau, J. (1984, p.11). In *Islands at the Edge – Preserving the Queen Charlotte Islands.* Vancouver, B.C.: Islands Protection Society/ Douglas and McIntyre.

19. Merlen, G. (1998, p.7). Human waves over Galapagos. *Islander*, 5, pp.2–9.

20. Lu, F., Valdiva, G., & Wolford, W. (2013). Social dimensions of "nature at risk" in the Galapagos Islands, Ecuador. *Conservation & Society*, 11(1), pp.83–95. Retrieved from http://www.conservatio-nandsociety.org/article.asp?issn=0972-4923;year=2013;volume=11;issue=1;spage=83;epage=95;aulast=Lu

21. Corcut, J. (2010, September 3). Forest fires in Madeira put future of Europe's rarest seabird under threat. *The Independent*, p.26.

22. Than, K. (2010, September 26). Drug-filled mice airdropped over Guam to kill snakes. *National Geographic Daily News*. Retrieved from http://news.nationalgeographic.com/news/2010/09/100924-science-animals-guam-brown-tree-snakes-mouse-tylenol/

Synge, H. (1990, September 28). Fencing with extinction. *The Guardian*, p.35.

23. Ascension: The island where nothing makes sense. (2016, April 19). *BBC News Magazine*. Retrieved from http://www.bbc.com/news/magazine-36076411

24. De la Mare, W. (1930). *Desert Islands and Robinson Crusoe*. London: Faber & Faber.

Kingsland, G. (1984). *The Islander: The Man who Wanted to be Robinson Crusoe*. London: New English Library.

Leslie, E.E. (1988). *Desperate Journeys, Abandoned Souls*. Boston, MA: Houghton Mifflin.

25. Watson, T. (2016, April 19). 80 rats exploded into 100,000 by avoiding poison. *National Geographic*. Retrieved from http://news.nationalgeographic.com/2016/04/160419-rats-exploded-poison-henderson-island/

26. Hunt, S. (2013, March 25). South Georgia leads where Britain may follow with mass reindeer cull. *The Independent*, p.25.

Leader-Williams, N. & Walton, D. (1989, February 11). The isle and the pussycat. *New Scientist*, pp.48–51.

O'Connor, S. (2015, June 25). Rare birds return to remote South Georgia island after successful rat eradication programme. *The*

References

Independent. Retrieved from http://www.independent.co.uk/environment/nature/rare-birds-return-to-remote-south-georgia-island-after-successful-rat-eradication-programme-10345864.html

27. Martinez, A.R. (2012). *Battle at the Edge of Eden.* Washington DC: The Atlantic Books. (Kindle Edition).

28. MacLean, C. (1972). *Island on the Edge of the World; Utopian St Kilda and its Passing.* London: Tom Stacey.

29. Cooper, D. (1985, p.203). *The Road to Mingulay.* London: Routledge & Kegan Paul.

30. Stride, P. (2008, April). St Kilda, the neonatal tetanus tragedy of the nineteenth century. *Journal of the Royal College of Physicians Edinburgh,* 38(1), pp.70–77. Retrieved from https://www.ncbi.nlm.nih.gov/pubmed/19069042

31. Saint Kilda – World Heritage Site Nomination Document. Retrieved from http://www.kilda.org.uk/kildanomdoc/topframe6.htm

32. MacLean (1972).

33. Nicolson (2002, p.269).
    Steele, T. (1975). *The Life and Death of St Kilda.* London: Fontana.

34. Anderson, I.F. (1930, July, p.273). The evacuation of St Kilda. *The Scots Magazine,* pp.265–273.

Chapter Four: The Economics of Vulnerability

The challenges facing "development" in small island economies receive critical examination in Dommen & Hein's *States, Microstates and Islands* (1985). I am particularly indebted to Polly Pattullo (1996a; 1996b) for her work on tourism. Mary Nazal (2005) provides valuable insights into legal aspects of environmental destruction on Nauru, and she has led me to other key sources. As ever, articles by Kathy Marks (2004a; 2008) and John Ezard (1993) have been thoughtful and informative. Pearl Binder's long-standing book on the "trials of the Ocean Islanders" (1977) is an unbridled account of injustice.

1. Brookfield, H.C. (1990). *An Approach to Islands.* In W. Beller, P. D'Ayala & P. Hein (Eds.). *Sustainable Development and Environmental Development of Small Islands.* Man and the Biosphere Series, Vol. 5. Paris and Carnforth, U.K: UNESCO/Parthenon.

2. Smith, H. (1995, July 31). Greece appeals to the weary and the Byronic to save its isles. *The Guardian,* p.7.
   Smith, H. (2008, October 17). Greek islands. *The Guardian,* p.8.

3. Lowenthal, D. (1992). *Small Tropical Islands: A General Overview.* In H.M. Hintjens & M. Newitt (Eds.), *The Political Economy of Small Tropical Islands,* pp.19–29. Exeter, U.K.: University of Exeter Press.

The Island

4. Dommen, E.C. & Hein, P.L. (1985). *Foreign Trade in Goods and Services: The Dominant Activity of Small Island Economies.* In E. Dommen & P. Hein (Eds.), *States, Microstates and Islands* (pp.152–184). London: Croom Helm.

    Wace (1980). *Exploitation of the Advantages of Remoteness and Isolation in the Economic Development of Pacific Islands.* In R.T. Shand (Ed.), *The Island States of the Pacific and Indian Oceans: Anatomy of Development* (pp.3–20, 87–118). Canberra, Australia: Development Studies Centre Monograph No.23, The Australian National University.

5. Pascal, J. (1997, November 1). Islands awash with ill-gotten gains. *The Guardian*, p.21.

    Taylor, J. (2008a, December 11). European feudalism finally ends. *The Independent*, pp.14–15.

6. Table adapted from Dommen & Hein (1985).

7. Dolman, A.J. (1985). *Paradise Lost? The Past Performance and Future Prospects of Small Island Developing Countries.* In E. Dommen & P. Hein (Eds.), *States, Microstates and Islands* (pp.40–69). London: Croom Helm.

8. Wace (1980).

9. Dommen, E.C. (1980, p.931). Some distinguishing features of island states. *World Development*, 8(12), pp.931–943.

10. World Bank (2012, December 12). *World Development Indicators Data Bank.* Retrieved from http://databank.worldbank.org/data/reports.aspx?source=world-development-indicators

11. Bass, S. & Dalal-Clayton, B. (1995). *Small Island States and Sustainable Development: Strategic Issues and Experience.* London: Environmental Planning Group, International Institute of Environment and Development, Environmental Planning Issues No. 8.

    Earth Summit: Programme of Action for Small Island States. (Bridgetown, Barbados, April 26–May 6, 1994). New York: United Nations Department of Public Information.

12. Binder, P. (1977). *Treasure Islands; the Trials of the Ocean Islanders.* Tiptree and London: Blond & Briggs.

13. The resettlement of the Banabians in Rabi, Fiji. (2011, May–June). *Bulla Bulletin.* Retrieved from http://www.methodist.org.uk/downloads/wcr-julia-edwards-may-june2011.pdf

    Binder (1977).

    Fairburn, T.L.J. (1985). *Island Economics. Studies from the South Pacific.* Suva, Fiji: University of the South Pacific, Institute of Pacific Studies.

14. Gowdy, J. & McDaniel, C.N. (2000). *Paradise for Sale. A Parable of Nature.* Oakland, CA: The University of California Press.

    Nazal, M. (2005, April). Nauru: An Environment Destroyed and International Law. *Law and Development.* Retrieved from http://www.lawanddevelopment.org/docs/nauru.pdf

References

15. Gowdy *et al.* (2000), p.3.
    Weeramantry, C. (1992), *Damage Under International Trusteeship.*
    Melbourne, VIC: Oxford University Press.
16. Nazal (2005).
17. Ezard, J. (1993, June 19–20, p.25). A drop in the ocean. *The
    Guardian Outlook*, p.25.
18. Marks, K. (2004a, April 18). Clouds over paradise as island of
    Nauru sinks into bankruptcy. *The Independent*. Retrieved from
    http://www.independent.co.uk/news/world/australasia/clouds-
    over-paradise-as-island-of-nauru-sinks-into-bankruptcy-560425.
    html
    Marks, K. (2008, February 21). South Pacific tragedy. *The Inde-
    pendent*, pp.24–25.
19. Nauru, Jurisdiction Spotlight. *Streber Weekly.* Retrieved from
    https://www.streber.st/2016/04/jurisdiction-spotlight-nauru/
20. Australian officials "paid people smugglers" – Amnesty. (2015,
    October, 28). *BBC News*. Retrieved from http://www.bbc.co.uk/
    news/world-australia-34654629
    Australia: Appalling abuse, neglect of refugees on Nauru. (2016,
    August 2). Amnesty International. Retrieved from https://www.
    hrw.org/news/2016/08/02/australia-appalling-abuse-neglect-
    refugees-nauru
    Marks, K. (2007b, December 11, p.26). Australia scraps "Pacific
    Solution" for refugees. *The Independent*, p.26.
    Marks, K. (2013, July 25). Claims of extreme suffering in refugee
    camps. *The Independent*, p.19.
    Sachs, A. & Naidoo, I. (1982). *Island in Chains*. Harmondsworth,
    U.K.: Penguin.
21. Hume, T. (2012, February 14). One is a destitute microstate. *The
    Independent*, pp.38–39.
22. Gowdy *et al.* (2000).
    Nazal (2005).
    Weeramantry (1992).
23. Dommen & Hein (1985).
    Pattullo, P. (1996a). *Last Resorts: The Cost of Tourism in the Caribbean.*
    London: Cassell.
24. Pattullo, P. (1996b, p.16). Trouble in paradise. *Islander*, (2),
    pp.16–17.
25. Lowenthal (1992).
26. Butler, R.W. (1986). The concept of the tourist area life-cycle:
    Implications for the management of resources, *Canadian Geogra-
    pher*, 24(1), pp.5–12.
27. Pattullo (1996a).
28. Foley, S. (2010, May 31). UK faces revolt in the Caribbean. *The
    Independent*, pp.6–7.

Marlow, B. (2013, April 14). Osborne cracks down on Caymans. *The Times Business Section*, pp.1–2.

29. Hodgson, M. (2002, June 11, p.7). Trouble in paradise. *The Guardian*, pp.6–7.

30. Alvarez, H. (2015, October 12). The island where men are disappearing. *BBC News*. Retrieved from http://www.bbc.co.uk/news/magazine-34487450

31. Fletcher, M. (2014, November 22). Pond Inlet: The Inuit's new struggle for survival. *The Telegraph*. Retrieved from http://www.telegraph.co.uk/news/earth/11240949/Pond-Inlet-the-Inuits-new-struggle-for-survival.html

32. Fletcher (2014).

   McKie, R. (2016, August 21). Inuit fear they will be overwhelmed as "extinction tourism" descends on the Arctic. *The Guardian*. Retrieved from https://www.theguardian.com/world/2016/aug/20/inuit-arctic-ecosystem-extinction-tourism-crystal-serenity

   Mission accomplished: Crystal Serenity completes 32-day North-West Passage journey. (2016, September 16). *Business Wire*. Retrieved from http://www.businesswire.com/news/home/20160916005705/en/Mission-Accomplished-Crystal-Serenity-Completes-32-Day-Northwest)

33. Marks (2008, p.5).

Chapter Five: Political Dependence and Turbulence

As in Chapter Four, I have found the collection of pieces in Dommen & Heim (1985) particularly helpful. Simon Winchester's *Outposts* (1985) has inspired my searches among the mostly island remnants of the British Empire.

1. Foote, T. (1986). How to keep the 20th century mostly at bay. *Smithsonian*, 17(2), pp.93–105.

   McCormick, D. (1974a). *The Islands of England and Wales*. Reading, U.K.: Osprey.

   MacLean, C. (1972). *Island on the Edge of the World; Utopian St Kilda and its Passing*. London: Tom Stacey.

   Schalansky, J. (2012). *Pocket Atlas of Remote Islands: Fifty Islands I Have Never Set Foot on and Never Will*. (C. Lo, Trans.). London: Penguin Books.

2. Howden, D. (2005, August 9). At the edge of the world. *The Independent*, pp.12–13.

   Michener, J.A. & Grove Day, A. (1957). *Rascals in Paradise*. New York: Ransom House.

   Porteous, J.D. (1981). *The Modernisation of Easter Island*. Victoria,

References

BC: Western Geographical Series Volume 19, Department of Geography, University of Victoria.

Rogers, R.A. (1926). *The Lonely Island*. London: George Allen & Unwin.

3. Lummis, T. (2000). *Life and Death in Eden. Pitcairn Island and The Bounty Mutineers*. London: Phoenix.

4. Taylor, J. (2008b, December 12). Sark feels wrath of the Barclay brothers. *The Independent*, p.5.

5. Marks, K. (2011, September 22). Scottish clan that wants its tropical paradise returned. *The Independent*, pp.34–35.

6. Doumenge, F. (1985, p.102). *The Viability of Small Intertropical Islands*. In E. Dommen & P. Hein (Eds.), *States, Microstates and Islands* (pp.70–118). London: Croom Helm.

7. McElroy, J. (1998). A propensity for dependence. *Islander*, 5, p.33.

8. Winchester, S. (1985). *Outposts*. London: Hodder & Stoughton.

9. Armesto, F.F. (2012, March 31, p.39). Thirty years on, the British still cannot admit the truth about the Falklands. *The Independent*, p.39.

    Bawden, T. (2012, March 16). Oil: Argentina threatens to sue Falklands drillers. *The Independent*, p.58.

10. Livingstone, G. (2012, January 22, p.38). Falklanders: We are the luckiest working-class people in the world. *The Independent*, p.38.

    Browne, A. (2002, March 17). Oh! What a lovely war. *The Observer Magazine*, pp.20–27.

11. Winchester, S. (1985, p.308).

12. Thomson, A. (1997, April 23–29, p.17). St Helena is simmering in exile. *The Weekly Telegraph*, p.17.

13. Fountain, N. (1999, October 23). Lost in space. *The Guardian*, p.6.

14. Background to Fiji's four coups (2006, December 8). *BBC News*. Retrieved from http://news.bbc.co.uk/1/hi/world/asia-pacific/6209486.stm

    Wright, O. (2013, September 28). Tourism, the Maldives' lifeblood, threatens to shut down over election. *The Independent*, p.25.

15. Bobin, F. (2012, March 2). India damps down Maldives tensions. *The Guardian Weekly*, p.12.

    Carrell, S. (2012, December 18). Officers face allegations of repeated civil rights abuses since coup. *The Guardian*, p.8.

Chapter Six: No Value But That of Location

I am grateful to Kevin Crossley-Holland (1972) for his graphic descriptions of abandoned militarised landscapes, and to Jerry Mander (1992) and Ellis & Blair (1986) for their critical accounts of the tragedy of American nuclear testing on Pacific islands.

The Island

Simon Winchester's work on the British island "outposts" of empire (1985) has again proved insightful, and I have drawn too on his graphic accounts of Pacific "dystopia". The writing of John Pilger (2006) and his related film, *Stealing a Nation* (Curtis, 2004), forms the basis for my account of the British government's clearance of the Chagos Islands.

1. Stummer, R. (1998, August 15). Once were warriors. *The Guardian Weekend*, pp.28–31.
2. Semple, E. (1911). *Influences of Geographic Environment*. New York: Henry Holt.
   Thomas, L. (1983). *A World of Islands.* London: Michael Joseph/Rainbird.
3. Porteous, J.D. (1981, p.30). *The Modernisation of Easter Island*. Victoria, BC: Western Geographical Series Volume 19, Department of Geography, University of Victoria.
4. Winchester, S. (1992). *The Pacific*. London: Arrow/Random Century.
5. Popham, P. (2012, September 5). The islands that divide superpowers. *The Independent*. Retrieved from http://www.independent.co.uk/news/world/asia/the-islands-that-divide-superpowers-8107033.html
6. Beech, H. (2016, May 23). What's new on China's artificial islands in the South China Sea? Basketball Courts. *Time News*. Retrieved from http://time.com/4341510/south-china-sea-artificial-islands/
   South China Sea: Satellite photos "show weapons built on islands". (2016, December 15). *BBC News*. Retrieved from http://www.bbc.co.uk/news/world-asia-38319253
7. Humphreys, A. (2012, April 11). New proposal would see Hans Island split equally between Canada and Denmark. *National Post News.* Retrieved from http://news.nationalpost.com/news/canada/new-proposal-would-see-hans-island-split-equally-between-canada-and-denmark
   Usborne, D. (2005, August 24). Canada makes show of force. *The Independent*, p.24.
8. Genttleman, J. (2009, August 23, p.20). Lake slum caught in Africa's fish war. *Scotland on Sunday*, p.20.
   Howden, D. (2009, March 23, p.22). Big trouble on small island. *The Independent,* pp.22–23.
9. Goodman, F.R.; Hargreaves, J.; Matthews, N. & Smith, B.J.N. (1978). *British Kayak Expedition Cape Horn. Official Expedition Report*. (Private Publication).
10. How Antarctic bases went from wooden huts to sci-fi chic. (2017,

References

January 13). *BBC News*. Retrieved from http://www.bbc.co.uk/news/magazine-38574003

Why do so many nations want a piece of Antarctica? (2014, June 20). *BBC News Magazine*. Retrieved from http://www.bbc.co.uk/news/magazine-27910375

11. Porteous (1981).

12. Munch, P. (1971). *Crisis in Utopia; The Ordeal of Tristan da Cunha*. New York: Thomas Y. Cromwell.

    Winchester, S. (1985). *Outposts*. London: Hodder & Stoughton.

13. Oliver, D.L. (1962, p.382). *The Pacific Islands*. Cambridge, MA: Harvard University Press.

14. Crossley-Holland, K. (1972). *Pieces of Land*. London: Victor Gollancz.

    Hetherington, P. (1990, April 25). Doubts raised over anthrax island. *The Guardian*, p.2.

    Mitchell, I. (2001). *Isles of the West*. Edinburgh: Birlinn.

    McCormick, D. (1974c). *The Islands of Scotland*. Reading, U.K.: Osprey.

15. Close, F. (1990). *Too Hot to Handle: The Race for Cold Fusion*. London: W.H. Allen.

16. Zinn, C. (1995, October 12). Lethal rabbit virus escapes from lab. *The Guardian*, p.15.

17. Crossley-Holland (1972, pp.32–33).

18. Crossley-Holland (1972, p.170).

19. Carrell, S. (1993, October 15). Isles' fears as missile base faces threat. *The Scotsman*, p.1.

    The Road to Isles Disasters (Editorial). (1993, October 15). *The Scotsman*, p.14.

20. Baker, N. (2008). *Human Smoke: The Beginnings of World War II, and the End of Civilisation*. London: Simon & Schuster.

21. Mander, J. (1992, p.346). *In the Absence of the Sacred*. San Francisco: The Sierra Club.

22. Price, W. (1966, p.66). *America's Paradise Lost*. New York: J. Day.

    Ellis, W.S. & Blair, J.P. (1986). Bikini – A Way of Life Lost. *National Geographic*, 169, pp.810–834.

23. Ellis & Blair (1986, pp.825, 828).

24. McIntosh, M. (1987). *Arms Across the Pacific*. London: Frances Pinter.

25. Chalmers, J. (1995, July 28, p.9). Paradise lost to nuclear mayhem. *The Guardian*, p.9.

26. Duval Smith, A. (1995, September 8, p.2). Fallout in the South Pacific. *The Guardian*, p.2.

27. Cited in Sengupta, K. (2009, June 6, p.26). Nuclear veterans win compensation case. *The Independent*, p.26.

28. McCarthy, M. (2010a, February 10, p.18). Man vs Marine. *The Independent*, pp.18–19.

29. Cited in Curtis, M. (2004, p.4). *Stealing a Nation*. Based on "A Special Report by John Pilger", Granada/ITV, transmitted October 2004. Colchester, U.K.: ITV.
30. Winchester, S. (1985, pp.39, 37).
31. Pilger, J. (2006, May 29). Out of Eden. *The Guardian*, pp.6–11.
32. Research Briefing Files, Parliament, U.K. Retrieved from http://www. researchbriefings.files.parliament.uk/documents.SN004463.pdf
    Tweedie, N. (2006, May 12). Britain shamed as exiles of the Chagos Islands win the right to go home. *The Telegraph*. Retrieved from http:// www.telegraph.co.uk/news/uknews/4200066/Britain-shamed-as-exiles-of-the-Chagos-Islands-win-the-right-to-go-home.html
33. Winchester (1985, p.38).
34. Calzonetti, C. (2012, 23 July). Council on Foreign Relations. FAQ about the ICC. Retrieved from http://www.cfr.org/ courts-and-tribunals/frequently-asked-questions-international-criminal-court/p8981
35. Cited in Pilger (2006, p.9).
36. Cited in McCarthy, M. (2010b, April 2). Preserved: Britain's "barrier reef". *The Independent*, pp.2–3.
37. Bowcott, O. (2016, November 17). Mauritius threatens to take Chagos Islands row to UN. *The Guardian*. Retrieved from https:// www.theguardian.com/world/2016/nov/17/mauritius-threatens-to-take-chagos-islands-row-to-un-court
38. HMG floats proposal for marine reserve covering Chagos Archipelago (2009). *Wikileaks*. Retrieved from http://www.telegraph. co.uk/news/wikileaks-files/london-wikileaks/8305246/ HMG-FLOATS-PROPOSAL-FOR-MARINE-RESERVE-COVERING-THE-CHAGOS-ARCHIPELAGO-BRITISH-INDIAN-OCEAN-TERRITORY.html

Chapter Seven: Paradise and Purgatory

As in Chapter Six, I have found the writing of Simon Winchester (1985; 1990) a very useful springboard. Bengt Danielsson's *The Happy Island* (1952) presents a window into an islander "paradise", whilst Gavan Daws (1980) locates Melville, Stevenson and Gaugin as outsiders in search of something similar. Adam Nicolson (2002) from his home in the Shiant Islands is well placed to describe just how close heaven and hell can be.

1. Cited in Morison, S.E. (1971, p.4). *The European Discovery of America; The Northern Voyages AD 500–1600*. New York: Oxford University Press.
2. Cited in Lockley, R.M. (1957, p.19). *In Praise of Islands*. London:

References

Frederick Mueler.

Ashe, G. (1962). *Land to the West*. London: Collins.

Wellard, J. (1975). *The Search for Lost Worlds*. London: Pan.

3. Higginson, T.W. (1898, p.92). *Tales of the Enchanted Isles of the Atlantic*. Great Neck, NY: Core Collection.

4. Tuan, Y-F. (1974, p.118). *Topophilia; A Study of Environmental Perception, Attitudes and Values*. Englewood Cliffs, NJ: Prentice-Hall.

5. Palmer R. & Cuffari, R. (1975). *A Dictionary of Imaginary Places*. New York: Henry Z. Walck.

6. Cited in Welch, H. (1966, p.92). *Taoism – The Parting of the Ways*. Boston, MA: Beacon.

7. Babcock, W.H. (1922). *Legendary Islands of the Atlantic: A Study in Medieval Geography*. New York: American Geographical Society.
Morison, S.E. (1971). *The European Discovery of America; The Northern Voyages AD 500–1600*. New York: Oxford University Press.
Sauer, C.O. (1968). *Northern Mists*. Berkeley and Los Angeles, CA: University of California Press.

8. Cited in Daws, G. (1980, p.4). *A Dream of Islands. Voyages of Self-Discovery in the South Seas*. New York and London: Norton.

9. Cited in Daws (1980, p.225).
Lummis (2000).

10. Cited in Daws (1980, p.212).

11. Marr, D. (1992, p.381). *Patrick White – A Life*. London: Vintage.
Brown, E. (2005). Fraser, Eliza Anne (1798–1858). *Australian Dictionary of Biography*. National Centre of Biography, Australian National University. Retrieved from http://adb.anu.edu.au/biography/fraser-eliza-anne-12929.

12. Bullen, F. (1944, p.189). *The Cruise of the "Catchalot". Round the World After Sperm Whales*. Harmondsworth, U.K.: Penguin.

13. Heyerdahl, T. (1963, pp.196–7). *The Kon-Tiki Expedition*. Harmondsworth, U.K.: Penguin.

14. Danielsson, B. (1952, pp.9–10). *The Happy Island*. London: George Allen & Unwin.

15. Lawson, D. (2007, July 6). From Pentecost Island to modern Britain. *The Independent*, p.33.

16. Nicolson, A. (2002, p.162). *Sea Room*. London: Harper Collins.

17. Johnson, S. & Boswell, J. (1924 ed. p.59). *A Journey Through the Western Islands;* and *The Journal of a Tour to the Hebrides*. R.W. Chapman (Ed.). London: Oxford University Press.

18. Flower, R. (1945, p.36). *The Western Island, or The Great Blasket*. New York: Oxford University Press.

19. Lavelle, D. (1976). *Skellig: Island Outpost of Europe*. Dublin: The O'Brien Press.
McCormick, D. (1974b). *The Islands of Ireland*. Reading, U.K.: Osprey.

The Island

McCormick, D. (1974c). *The Islands of Scotland*. Reading, U.K.: Osprey.

Moore, B. (1972). *Catholics*. Toronto: McClelland & Stuart.

20. Fish, G. (1982). *Dreams of Freedom: Bella Coola, Cape Scott, Sointula*. Victoria, BC: Provincial Archives of British Columbia.

    Hodgins, J. (1978). *The Invention of the World*. Scarborough, ON: New American Library of Canada (Signet Edition).

    Wilson, H.E. (1967). *Canada's False Prophet: The Notorious Brother Twelve*. Richmond Hill, ON: Simon & Schuster.

    Woodcock, G. (1960). *The Spirit Wrestlers: An Account of the Doukhobors*. New York: Meridian.

21. Guppy, S. (1985, p.55). "Ichthus". In *Another Sad Day on the Edge of the Empire* (pp.52–74). Lantzville, BC: Oolichan.

22. Raban, J. (1999, pp.338–9, p.339). *Passage to Juneau*. New York: Pantheon.

    Kennedy, L. (1991). *Coastal Villages of British Columbia*. Madeira Park, BC: Harbour Publications.

23. Porter, P.W. & Lukermann, F.E. (1976). *The Geography of Utopia*. In D. Lowenthal & M.J. Bowden (Eds.), *Geographies of the Mind. Essays in Historical Geosophy* (pp. 197–223). New York: Oxford University Press.

24. Semple, E. (1911). *Influences of Geographic Environment*. New York: Henry Holt.

25. Leslie, E.E. (1988). *Desperate Journeys, Abandoned Souls*. Boston, MA: Houghton Mifflin.

    Ramsay, R.H. (1972). *No Longer on the Map – Discovering Places that Never Were*. New York: Viking

26. Schalansky, J. (2012). *Pocket Atlas of Remote Islands: Fifty Islands I Have Never Set Foot on and Never Will*. (C. Lo, Trans.). London: Penguin Books.

    Banning, G.H. (1925). *In Mexican Waters*. London: M. Hopkinson.

27. Winchester, S. (1985, p.116). *Outposts*. London: Hodder & Stoughton.

28. Winchester, S. (1992, pp.182, 187). *The Pacific*. London: Arrow/Random Century.

29. Muir, J. (1917, pp.36–7). *The Cruise of the Corwin; Journal of the Arctic Expedition of 1881 in Search of De Long and the Jeanette*. Boston and New York: Houghton Mifflin.

    Matsen, B. (1986). The Aleutians: Dark current, dark land, resilient people. *Oceans*, 19(1), pp.34–43, 71.

30. Cited in Birds of a feather. (2010, September 14, p.31). *The Independent*, pp.30–31.

31. Cited in Schalansky (2012, p.21).

32. Cited in Simmons, J.C. (1998, p.36). *Castaway in Paradise*. Lanham, MD: Sheridan House.

References

33. Brooks, L. (1999, January 16). Hope warms the grim edge of Paradise. *The Guardian,* p.17.
    Cornwell, T. (1993, June 13). Island of the damned. *The Observer Magazine*, pp. 41–43.

34. Lichfield, J. (2012, November 16). The most murderous place in Europe. *The Independent*, p.37.
    Lichfield, J. (2013b, June 29). Finally the Tour rolls in to Corsica – but is it safe? *The Independent*, p.30.

35. Barbour, I.G. (1977). *Technology, Environment and Human Values*. New York: Praeger.
    Walsh, C. (1966). *From Utopia to Nightmare*. New York: Harper & Row.

36. Brook, M. (2012). The tyrant of Clipperton Island. Retrieved from http://www.damninteresting.com/the-tyrant-clipperton-island/
    Lichfield, J. (2007, February 5). Slavery and its legacy. *The Independent*, pp.24–25.

37. Golding, W. (1956). *Pincher Martin*. London: Faber & Faber.

38. Sachs, A. & Naidoo, I. (1982, p.225). *Island in Chains*. Harmondsworth, U.K.: Penguin.
    De Villiers, S.A. (1971). *Robben Island – Out of Reach, Out of Mind*. Cape Town: C. Struik.

39. Hughes, R. (1987, p.457). *The Fatal Shore*. London: Pan.

40. Hughes (1987).

41. Sachs & Naidoo (1982).

42. Starkins, E. (1976). *A BC Leper Colony*. In H. White (Ed.), *Raincoast Chronicles First Five* (p.51). Madeira Park, BC: Harbour.

43. Bowering, M. (1989, pp.178, 145). *To All Appearances a Lady*. Toronto: Random House.
    Bristowe, W.S. (1969). *A Book of Islands*. London: G. Bell.

44. Bristowe, W.S. (1969). *A Book of Islands*. London: G. Bell.

45. Semple, E. (1911). *Influences of Geographic Environment*. New York: Henry Holt.

46. Hughes (1987, p.484).

47. Heikell, R. (2002, p.317). *Italian Waters Pilot*. St Ives, U.K.: Imray & Laurie Norie & Wilson.

48. Smith, H. (2008, October 17, p.8). Greek islands. *The Guardian*, p.8.

49. Cited in Merritt, J. (1989, September 10, p.7). Europe's guilty secret. *The Observer*, pp.1, 7.

50. Cited in The naked and the damned. (1989, September 10, p.17). *The Observer*, p.17.
    Hislop, V. (2006). *The Island*. London: Review/Hodder Headline.

51. Lonsdale, S. (1992). Children in chains on island of the damned. *The Observer*, November 8, p.13.

Chapter Eight: An Island Mentality

John Fowles (1978; 1981) and Adam Nicolson (2002) have quite particular views on islandness that have catalysed my ideas. I am heavily indebted to Trevor Lummis (2000) for his history of Pitcairn Island, and to Dea Birkett (1998) for her insightful account of living on the island. Without Somerset Maugham's "German Harry" and D.H. Lawrence's "The Man Who Loved Islands" our exploration of an island mentality would be incomplete.

1. Bates, H.E. (1972). *The World in Ripeness. An Autobiography, Vol. 3*. London: Michael Joseph.
2. Theroux, P. (1992). *Happy Isles of Oceania*. London: Penguin.
3. Nicolson, A. (2002, pp.13, 141). *Sea Room*. London: Harper Collins.
4. Nicolson (2002, p.141).
5. Cited in Conover, D. (1980, p.135). *One Man's Island*. Markham, ON: Paperjacks.
6. Hamsun, K. (1977, p.104). *The Wanderer*. London: Picador.
7. Somers, D. (1994, p.159). "The stone boat". In *At the Rising of the Moon*, pp.155–168. London: Baton Wicks.
8. Fowles, J. & Godwin, F. (1978, pp.78, 11–12). *Islands*. London: Jonathan Cape.
9. Fowles, J. (1981, viii). *Introduction*. In G. E. Edwards, *The Book of Ebenezer Le Page*. New York: Alfred A. Knopf.
10. Frater, A. (2004). *Tales from the Torrid Zone*. London: Picador
11. Guterson, D. (1996, pp.385–6). *Snow Falling on Cedars*. London: Bloomsbury.
12. Birkett, D. (1998, pp.167–8). *Serpent in Paradise*. London: Picador.
13. Conover (1980).
    Micham, A. (1984). *Offshore Islands of Nova Scotia and New Brunswick*. Huntsport, NS: Lancelot.
14. Birkett (1998, p.60).
15. Daws, G. (1980, p.16). *A Dream of Islands. Voyages of Self-Discovery in the South Seas*. New York and London: Norton.
16. Cited in Birkett (1998, pp.228–9).
17. Cited in Lummis, T. (2000, p.204). *Life and Death in Eden. Pitcairn Island and The Bounty Mutineers*. London: Phoenix.
    Bennett, F. D. (1840). *Narrative of a Whaling Voyage 1833–1836*. London: Richard Bentley.
18. Cited in Lummis (2000, p.227).
19. Winchester, S. (1985). *Outposts*. London: Hodder & Stoughton.
20. Birkett (1998, p.15).
21. Young, R.A. (1894). *Mutiny of the Bounty and the Story of Pitcairn Island, 1790–1894*. Oakland, CA: Pacific Press, pp.252–3. Cited in

References

Lummis (2000, p. 241).

22. Birkett (1998).

23. Birkett (1998, p.109).

24. Noonan, M. (1983, p.148). *A Different Drummer; The Story of E.J. Banfield, the Beachcomber of Dunk Island*. St Lucia, Australia: University of Queensland Press.

25. Banfield, E.J. (1968, p.35). *The Confessions of a Beachcomber*. London: Angus and Robertson.

26. Chamberlain, B. (1987, p.225). *Tide Race*. Bridgend, U.K.: L. Seren.

27. Parker, T. (1986). *Lighthouse*. London: Eland.

28. Somerset Maugham, W. (1954). "German Harry". In *The World Over, The Collected Stories, Volume 2*, pp.81–84. London: Heineman (Reprint Society Edition).
    Lawrence, D.H. (1976 ed.). "The Man Who Loved Islands". In *The Complete Short Stories, Vol.3*, pp.722–746. Harmondsworth, U.K.: Penguin.

29. Somerset Maugham (1954, p.84).

30. Burchill, E. (1972). *Thursday Island Nurse*. Adelaide: Rigby.

31. Zetterland, T. (1975). Extract from *Den overkorda kangurun*. Retrieved from http://www.riverbendnelligen.com/ronbrand-tenglish.html

32. Nicolson (2002).

33. Nicolson (2002).

34. Tindall, G. (1991). *Countries of the Mind. The Meaning of Place to Writers*. London: The Hogarth Press.

Island Tales

1. Peachey, P. (2010, March 4). How a 12-year-old girl saved her Chilean island. *The Independent*, p.3.

2. Morris, S. (2010, May 10). British sailor saves Kanton islanders from starvation. *The Guardian*. Retrieved from https://www.theguardian.com/world/2010/may/10/british-yachtsman-kanton-island-resue

3. Smith, H. (1998, August 1). Blood spilled in ferry war. *The Guardian*, p.24.

4. Birkett, D. (1998). *Serpent in Paradise*. London: Picador.
   Shipwrecked on Oeno Island in 1858. Wreck of the "Wildwave": The Diary of Capt. Josiah N. Knowles. Retrieved from http://www.winthrop.dk/wildwave.html

5. Boffey, C. & Compston, E. (1997, October 5). Wildlife couple refuse to quit Caribbean isle. *The Sunday Telegraph*, p.19.

6. Anderson, S. (2010, December 21). Another attempt at locating

treasure buried on Robinson Crusoe Island. *Mercopress*. Retrieved from https://www.theguardian.com/world/2005/sep/26/chile. mainsection

Franklin, J. (2005, September 26). 600 barrels of loot found on Crusoe island. *The Guardian*. Retrieved from https://www. theguardian.com › World › Chile

7. Drummond, A. (1988, September 4). Lord Jim takes on Canberra. *The Observer*, p.25.

8. Davison, P. (2012, October 12). Major Roy Bates. The self-proclaimed "Prince of Sealand". Obituary. *The Independent*. Retrieved from http://www.independent.co.uk/news/obituaries/major-roy-bates-the-self-proclaimed-prince-of-sealand-8207988.html

9. Lummis, T. (2000). *Life and Death in Eden. Pitcairn Island and The Bounty Mutineers*. London: Phoenix.

10. McCormick, D. (1974a). *The Islands of England and Wales*. Reading, U.K.: Osprey.

11. Cooper, D. (1985). *The Road to Mingulay*. London: Routledge & Kegan Paul.

12. Thomas, L. (1983). *A World of Islands*. London: Michael Joseph/ Rainbird.

13. There are numerous accounts of this mystery: just enter "crew of the Sarah Joe" into an internet search engine.

14. Leslie, E.E. (1988). *Desperate Journeys, Abandoned Souls*. Boston, MA: Houghton Mifflin.

15. Coghlan, N. (2011). *Winter in Fireland*. Edmonton, AH: University of Alberta Press.

O'Shaughnessy, H. & Sohr, R. (1988, July 24). Pinochet plots exile course to Crusoe isle. *The Observer*, p.24.

Maps

The location map outlines are © d-maps.com and can be found at the urls below:

http://www.d-maps.com/carte.php?num_car=3258&lang=en
http://www.d-maps.com/carte.php?num_car=3200&lang=en
http://www.d-maps.com/carte.php?num_car=5864&lang=en
http://d-maps.com/carte.php?num_car=1384&lang=en
http://d-maps.com/carte.php?num_car=3122&lang=en
http://d-maps.com/carte.php?num_car=3246&lang=en
http://www.d-maps.com/carte.php?num_car=2554&lang=en

Index

Index

"Roderick the Imposter" 156
Rogers, Woods 41
Ross, Capt. John 28
Rousseau, Jean-Jacques 222, 275

"selling the nation" (Greek islands) 120
"stealing a nation" (Chagos) 209–16
"string of pearls" (Indian Ocean)
Ratcliffe Smiley 268

Saint Brendan 15, 22– 23, 29, 33
satire 16–17
seed banks 121
Selkirk, Alexander 41–2, 49, 56–7, 61, 109, 304
selling paradise 143, 151–2
species: invasive 98, 100, 108–10; eradication programmes 110–11; endangered 105–7; extinct 106
stamps 122–3, 125, 169
Stanbury, Bernard 57
Stevenson, Robert Louis 20, 152, 223
Swift, Jonathan 17–18

terror 247: *Pincher Martin* 247–8; Norfolk Island 295
Torres Strait (beachcombers/voluntary

maroons) 281–5
Toscanelli, Paolo 30
tourism 144–7; and crime 147–8, 150, 162; "extinction" 153; "global warming" 85; "paradise" 143, 151–2; "new frontiers" 152–4; "phantom" 153; legacy of colonialism (Caribbean) 146; economic dependency and wealth disparity 145–6; Butler Model 147
Travels of Sir John Mandeville 16

utopias: 6, 101, 235–6; *Islands of the Blessed, Fortunate Islands (Promised Land)* 22–24; *Island of Hope* 50;
United Nations Climate Summit Paris 2015 93

volcanoes 83–4, 88–91 *see also individual islands in Index of Islands*
Voyage of Bran 23

White, Patrick: *A Fringe of Leaves* 222–4
Wilkes Expedition 31

Index of Islands

Agothonisi: 254 asylum seekers
Agrihan: 88 evacuation
Aitutaki: 197 militarisation
Alamagan: 88 evacuation
Aldabra: 122 climate research; 210 ecology
Alderney: 200 military archaeology
Aleutians: 239 hyperbolic imagination; *see also individual islands*
Amchitka: 180 militarisation
Amorgos: 222
Anatahan 88 evacuation
Anchorage (Cook Islands): 21 hurricane; 54 voluntary marooning; 63-79 Tom Neale; 197 militarisation
Andaman: 36 chart errors
Anglesey: 306
Anguilla: 124 finance
Anjouan: 162 inter-island conflict
Antigua: 91 evacuees from Montserrat; 145,148 tourism and related crime; 158 political anachronisms, 179

geopolitics, 296-7
Antilia: 29 commercial subterfuge
Aran Islands: 24 Brasil; position doubtful; 234 monastic
Ascension: 40-1 marooning; 108-9 invasive species, 123 stamps;168-9 British 'outposts'; 195-6 geopolitics; 238-9 hyperbolic imagination
Atlantis: 12, 14–15 Plato *et al.*; 83 origins, fascination
Attu: 239 hyperbolic imagination; 180 militarisation
Azores: 23 *Brasil;* 83 *Atlantis;* 122 climate research; 221 *Antilia see also individual islands*

Bahamas: 23 Saint Brendan; 83 *Atlantis;* 145-6 tourism *see also individual islands*
Bahrain: 123 finance
Banaba (Ocean): 130-3 phosphate mining
Barbados: 145-6,148 tourism, related

335

Index of Islands

Index of Islands

The Island